DRAGONS IN ZOOLOGY, CRYPTOZOOLOGY, AND CULTURE

DR. KARL P. N. SHUKER

COACHWHIP PUBLICATIONS

Greenville, Ohio

DEDICATION

IN MEMORY of my good friend and fellow cryptozoologist Ivan Mackerle (1942-2013), who searched diligently for many elusive creatures, including such modern-day dragons as the tatzelworm and Mongolian death worm.

I am so sorry that you are no longer here, but I pray that you now have the answers to all of those fascinating mysteries that you pursued for so long and for which we who remain here still strive.

CONTENTS

ACKNOWLEDGEMENTS

I WISH TO THANK the following persons and institutions for their much-appreciated assistance and encouragement given to me during the preparation of this book, and in particular to those many talented artists who so generously made available and even specially prepared some of the beautiful illustrations included in it.

Silvana Pellegrini Adam, Chad Arment/ Coachwhip Publications, Lance Bradshaw, Markus Bühler, Pat Burroughs, the late Mark Chorvinsky, Thomas Finley, Miroslav Fišmeister, Prof. Roy P. Mackal, the late Ivan Mackerle, Tim Morris, Hodari Nundu, Andy Paciorek, William Rebsamen, the late Mary D. Shuker, Rebekah Sisk, Richard Svensson, Lars Thomas, Vorderasiatisches Museum Berlin, Anthony Wallis.

The author has sought permission for the use of all illustrations and substantial quotes known by him to be still in copyright. Any omissions brought to his attention will be rectified in future editions of this book.

INTRODUCTION

I am the old Dragon that is present everywhere on the face of the Earth,—father and mother, youthful and antique, weak yet powerful, visible and invisible, hard and soft, descending to the earth and ascending to the heavens, most high and most low—in me the order of nature is oftentimes inverted—I am dark and bright; I spring from the earth and I come out of heaven. I am well known yet a mere nothing.

Thomas Vaughan the Alchemist (1621-1666)

OF ALL THE countless legendary beasts that have been conjured forth from the seemingly limitless capacity of the human imagination, none can remotely compare with the dragon for its sheer diversity of form, its symbolic significance, and its cross-cultural presence. Dragons are everywhere—still glimpsed in the living, breathing beasts around us that inspired and engendered their birth in our far-distant ancestors' dreams, and nightmares; perennially encountered in the myriad of traditional myths and folklore woven into the fabric of every creed and culture around the world; and ever-visible within the innumerable outpourings of artistic creation that have graced and enhanced our species' existence across all temporal, political, social, and geographical boundaries.

Indeed, so diverse in form, behaviour, and occurrence is the dragon that it even defies any concise attempt at definition. Just what *is* a dragon?

In order, therefore, to avoid the innumerable specifics that delineate and multiply these fantasy forms, the best, most generalised definition that can be safely offered is that a dragon is a fabulous reptilian creature of global occurrence in traditional legends and lore, mostly malevolent in the West but often benevolent in the East, divisible into numerous subcategories based upon morphology and behaviour, and most commonly (though certainly not invariably) large in size. Some have limbs, some have wings, some have both, some have neither. Some breathe fire, some breath toxic vapour, some do neither.

The author and a reassuringly subdued dragon. (Dr. Karl Shuker)

WHEN DRAGONS WERE REAL

Today, there remains much discussion and dissension as to the origin of dragons—what inspired our global belief in these awe-inspiring monsters of mythology? Centuries ago, however, there was far less contention, for the simple reason that even the most educated of scholars and chroniclers, let alone the more credulous, uneducated laypeople, firmly believed in the reality of dragons.

Athanasius Kircher (1601/1602-1680) was a celebrated German Jesuit scholar, whose studies and numerous publications encompassed a vast range of subjects, including geology, theology, alchemy, biology, the hieroglyphics of ancient Egypt, medicine, and Oriental studies. In his geological publications, this highly learned researcher professed a profound belief in the existence of dragons, cataloguing many reports of several different categories of dragon. He even claimed that the fossilised bones of dinosaurs and other large prehistoric creatures sometimes unearthed during excavations were the mortal, petrified remains of lately-deceased dragons. And in answer to anyone who may not have shared his belief, he stated with uncompromising directness:

> There are reports of flying dragons from so many and such reliable witnesses, that anyone, I think, who denies their veracity must be himself completely mad.

Another leading believer in dragons was English cleric and bestiary compiler Edward Topsell (c.1572-1625). His two principal zoological works were *The Historie of Foure-Footed Beastes* (1607) and *The Historie of Serpents* (1608), which in 1658 were republished together as a massive 1100-page treatise entitled *The Historie of Foure-Footed Beastes and Serpents*.

DE DRACONE.

A page of dragon illustrations from Gesner's bestiary of 1658.

In it, he propounded at length his conviction that dragons existed, and in many different forms, citing numerous previous writers in support of his claims (but particularly Swiss scholar Conrad Gesner (1516-1565), whose own bestiary, *Historiae Animalium*, Topsell liberally referenced, as

well as ancient scholars such as Grevinus, Gillius, and Pierius among others):

> There be some dragons which have winges and no feete, some againe have both feete and wings, and some neither feete nor wings, but are onely distinguished from the common sort of Serpents by the combe growing upon their heads, and the beard under their cheeks . . .

In that short paragraph, Topsell succinctly summarised the traditional delineations of dragon form present in myth and legend that will be adhered to and explored in much greater detail within this present book.

Elsewhere in his tome, Topsell revealed:

> Saint Augustine saith, that dragons doe abide in deepe Caves and hollow places of the earth, and that some-times when they perceive moistnes in the ayre, they come out of theyr holes, and beating the ayre with their winges, as it were with the strokes of oares, they forsake the earth and flie aloft: which wings of theirs are of a skinny substance, and very voluble, and spreading themselves wide, according to the quantitie and largenesse of the dragons bodie.

Again, this encapsulates centuries of preserved lore as to the habitat and behaviour of these reptilian monsters, yet

Topsell presented it not as fable but rather as indisputable fact.

Once the bestiary compilers were replaced by scientifically-minded naturalists, however, it became evident that dragons were no more real than other fauna of the fabulous, such as unicorns, harpies, centaurs, and flying horses. No longer was it sufficient to quote and regurgitate the writings of bygone chroniclers in support of these creatures' existence. Now, physical, tangible evidence was required, but this was found to be sadly lacking in relation to dragons.

Where were their carcasses, their preserved skeletons, or even relics such as a mounted trophy head, or a taxiderm dragonet? As will be unmasked in the zoological sections of this present book, everything once deemed to have originated from a dragon was duly exposed to be from some other, much more prosaic animal instead—dragon bones and dragon teeth were the fossils of ancestral creatures that had died out millions of years ago, stuffed dragons were poorly-preserved crocodiles or cleverly-manufactured fakes using dried fishes or composites of several different animals, and so forth. After countless ages of domination and terror, the dragon was no more, simply because it had never been.

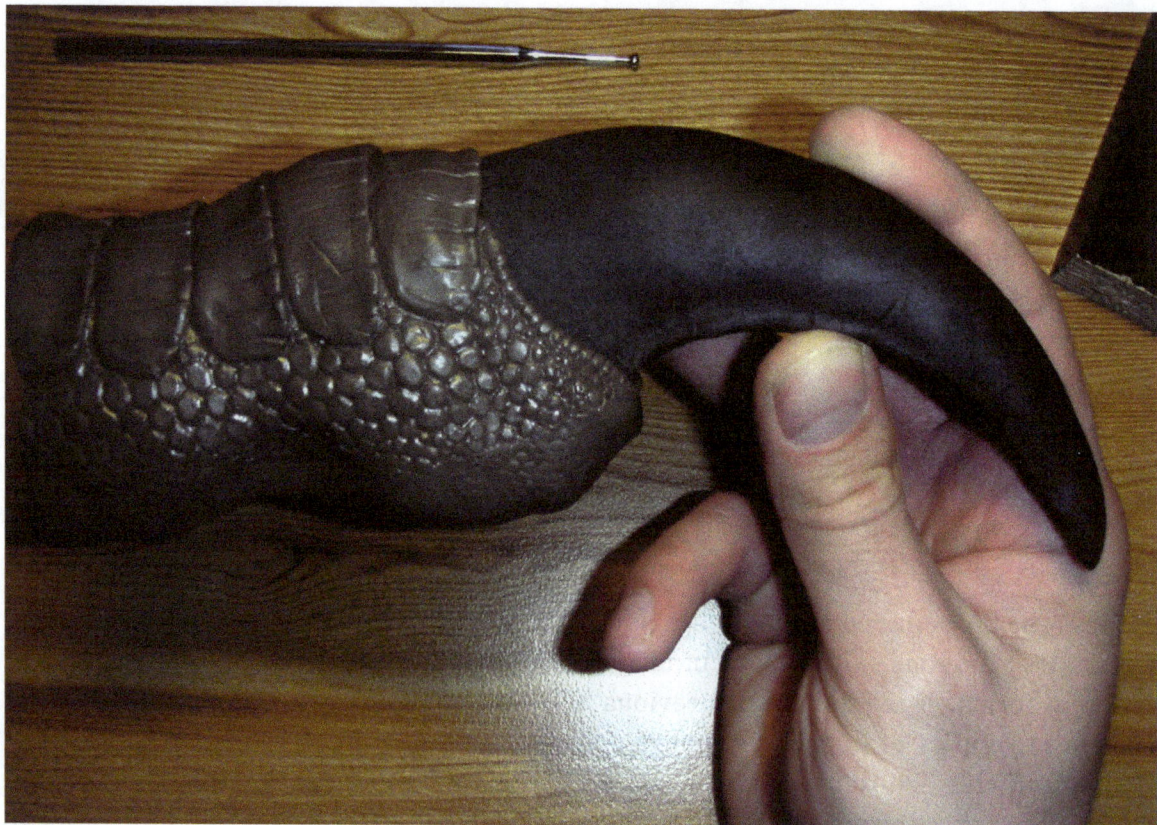

Model of a dragon's claw. (Markus Bühler)

Champ-like aquatic mystery beast—a modern-day water dragon? (William Rebsamen)

DRAGON-INSPIRING ANIMALS

So from where, and from what, had such widespread—indeed, worldwide—belief in dragons stemmed? Clearly, the dragon's origin, evolution, and persistence in our world is a fascinating subject to investigate, revealing and elucidating the unparalleled variety and complexity of dragon forms that have arisen through the ages in a startlingly close parallel to the natural evolution of real animal species, and which are explored within this present book in greater detail than within any previously-published work. Moreover, these many forms are linked in innumerable and often unexpected ways to their putative antecedents within the living world—because there can be no doubt that a major factor influencing the origin of the dragon is early humanity's observations and interactions with various distinctive and potentially inimical creatures of reality sharing our world.

Serpent dragons, for instance, undoubtedly owe their origin to encounters with

Dragon statue at a Buddhist temple in Luang Prabang, Laos.

some of the modern world's most sizeable species of snake. The large quadrupedal wingless dragons often faced down and dispatched by knights errant assuredly originated from exaggerated retellings of real-life encounters with crocodiles and giant lizards such as monitors. And pre-scientific attempts to elucidate fossils of pterodactyls no doubt inspired many a myth of winged dragons.

Also greatly deserving of detailed scrutiny, yet previously attracting surprisingly scant attention even by the most dedicated of draconophiles, is the plethora of modern-day reports concerning mysterious, unidentified creatures on land and in aquatic habitats that bear more than a passing resemblance to the dragons of fable. Do these elusive beasts truly exist, awaiting formal scientific discovery and description, or are they just distorted, half-remembered shadows and phantoms persisting from the legends and superstitions of our forefathers in long-bygone times?

The two-limbed serpentine tatzelworm of the Alps, for example, corresponds very closely indeed with folkloric descriptions of wingless two-limbed European dragons called lindorms. Long-necked lake monsters such as Nessie of Scotland's Loch Ness and Champ of Lake Champlain in North America irresistibly recall age-old tales of water dragons, as do various sea serpents and certain other marine mystery beasts. Fable or fact—non-existent monsters, or undiscovered dragons?

DRAGONS BY ASSOCIATION

Equally thought-provoking is how and why the dragon has become so intimately associated with our own species. For it has been interpreted in countless different ways by the nations of humanity across the globe, not just in the past but also in the present—and it will undoubtedly continue to be in the future too.

Of all legendary beasts, this multi-faceted monster of mythology is more than amply represented visually by artwork of every conceivable style, age, and category—from classic paintings, temple statues, tattoos, and digital art to cartoons, costumes, ornaments, heraldic emblems, and so much more. And the dragon's symbolic status in religion, dreams, alchemy, psychology, astrology, and other fields is as fascinating as it is complex. All of these compelling, thought-provoking subjects are extensively surveyed here.

Nor can anyone overlook the prominent role and immense, continuing popularity of this extraordinary beast in contemporary literature, music, films, computer games, and related media, which are also reviewed here, bringing this book's coverage completely up-to-date.

More dramatic still is the compelling prospect that our enduring fascination with dragons may even be intimately linked with preserved racial memories dating right back to the precarious survival of our species' most ancient ancestors in prehistoric times, when the world was dominated by real giant reptiles.

Consequently, with such a vast wealth of material to document and assess, each chapter in this present book chronicles and examines a very different aspect of the dragon's existence, with particular emphasis upon presenting the case from as wide a range of geographical regions as possible, in order to demonstrate just how divergent yet universal the dragon has become through time across the planet. Its contents split very conveniently into two readily divisible parts—the dragon as a zoological (and cryptozoological) entity, and the dragon as a cultural icon. And throughout the book, numerous examples of dragon myths, legends, and lore from around the world are recalled and retold.

THESE DIVERSE but equally captivating themes all provide ample confirmation that there is no sign whatsoever of waning interest for what must surely be the most vibrant, tenacious, and fascinating creature that has never existed—the dragon.

PART I:
DRAGONS IN ZOOLOGY AND CRYPTOZOOLOGY

CHAPTER I:
THE NATURAL HISTORY OF LOWER DRAGONS

OF ALL THE NUMEROUS mythological animals ever documented, none is more diverse in form and behaviour than the dragon. Indeed, during the development and flourishing of legends and myths worldwide, several distinct categories of dragon have been differentiated and named. Moreover, these collectively display a veritable evolutionary development, progressing from the most basic types to the most advanced ones in a manner that is analogous to the evolution of real creatures. Consequently, based upon their level of morphological and behavioural complexity, these varied dragon categories can in turn be grouped into two primary divisions—lower, simple, or basal dragons (predominantly serpentine in form with only—at most—a single pair of limbs and sometimes a pair of wings, and generally exhaling venomous breath); and higher, complex, or advanced dragons (quadrupedal in form with or without wings, and generally fire-breathing, plus, in certain

cases, an assortment of additional features, such as extra limbs, shape-shifting abilities, brow-embedded precious stones). This chapter surveys the world's very extensive array of lower dragons.

LIMBLESS WINGLESS DRAGONS— THE SERPENT DRAGONS

The term 'dragon' is widely believed to derive from the Greek word 'derkein', which translates as 'sharp-eyed', and seems to have been applied originally to a snake. When transliterated into Latin, it became 'draco'—'giant snake'. This is particularly apt because, morphologically, the basal dragon form was the serpent dragon. As its name suggests, in overall appearance this form, which was both limbless and wingless, resembled a huge snake or even a gigantic worm.

Indeed, some serpent dragons were actually referred to as worms, also known as wyrms (an Old English term for snakes), wurms, or orms—all of which are terms derived from the Norse 'ormr' or 'ormer', translating as 'dragon'. Unlike these humble, harmless beasts, however, and betraying its draconian status, a serpent dragon had a head that was horned, long-jawed, and very often bearded.

THE WORM

Just as there are several categories of dragon, equally there are several types of

A worm or wyrm, as depicted in Andy Paciorek's book *Strange Lands*. (Andy Paciorek)

serpent dragon, distinguished not only by form but also by geographical location or habitat. Many of Britain's dragons were of the worm variety, which was also recorded widely across northwestern Europe. Dragon-headed and generally of immense length, worms not only lacked limbs and wings but also one of the characteristics most readily linked with dragons—the ability to breathe fire. However, this did not make them any less deadly—far from it.

Instead of fire, worms emitted noxious clouds of poisonous gas that could lay waste to great swathes of countryside and decimate entire villages, or possessed a venomous bite of lethal propensity. In the case of a type of French worm known as the guivre, which inhabited forests, marshes, and sometimes rivers too, so virulent was its breath that it generated and spread all manner of deadly diseases, wiping out the inhabitants of every town or village that the guivre passed by.

Although worms did not exhale fire from their jaws, at least one version discharged fire from its tail. This unusual creature was the ejderha or evren of Turkey.

Worms could also kill by wrapping their huge coils around any potential antagonist, rather like real-life pythons and boas. They could even survive being chopped into several pieces, because the pieces swiftly reconnected with one another unless they were buried separately or burnt immediately.

An even odder characteristic of worms that occurs in many tales concerning them was their propensity for milk, which in some cases was such that worms were even encountered curled under cows, sucking their milk directly from their udders. Needless to say, the worms' venom meant that the cows rarely survived this unnatural milking, which was ruinous for their farmer owners.

The Italian boas, from Edward Topsell's *The Historie of Serpents,* 1608.

Perhaps the most notable udder-sucking worm was a huge Italian example known as the boas, which was first documented during the 1st Century AD in Pliny the Elder's *Natural History*. Often portrayed with a grossly distended body (due perhaps to the vast quantities of milk that it imbibed), the boas would actively pursue flocks of cattle and after attaching itself to the udders of a cow would proceed to suck it dry of milk, killing the unfortunate creature in the process, before moving onto the next. It even derived its name from 'Bos'—Latin for 'cow'—and its own name in turn gave rise to 'boa', used today as both the common and the scientific name for a specific taxonomic group of modern-day constricting snakes.

Speaking of which: another modern-day group of constricting snakes, the pythons, also derive their name from a European serpent dragon. Python was the earth dragon of Delphi, a limbless goliath that, depending upon which version of the myth is consulted, dwelt upon Mount Parnassus in Greece or lived in a deep chasm beneath the Delphic oracle, until slain by Apollo, the sun god. Similar serpent dragons elsewhere in Greece and also in Rome were often maintained in temples or caves, where they were worshipped by the local people as oracles or even as deities.

Very different from those was Greek mythology's shimmering-scaled, multi-headed serpent dragon Ladon, who guarded the golden apples in the Garden of the Hesperides (a trio of nymphs), and is often depicted coiled around these magical apples' tree. According to most versions of the myths concerning the twelve great labours of Heracles, when this famous hero was ordered to steal the golden apples he overcame Ladon, but did not kill it outright.

Another multi-headed serpent dragon was the South American ihuaivulu, which resided deep within the fiery heart of volcanoes, its burnished red and copper-hued scales perfectly complementing its incandescent abode. Unusually for serpent dragons, but not unexpected in view of this particular version's specific dwelling place, the ihuaivulu could breathe fire, spewing forth flames from the jaws of each of its seven heads. And in Chile's Araucanian region, native folktales tell of the cherrufe, a seven-headed serpent dragon that could not only breathe fire but also transform itself into other animal forms.

There are many memorable myths and legends concerning worms, which often include a cunning plan to destroy them, on account of their unnerving ability to recombine even when hacked into several pieces.

Perhaps the most famous British worm is the Lambton worm, which reputedly grew from a small black leech-like beast caught in the River Wear, County Durham, on Easter Sunday 1420 by John Lambton—the youthful, impious heir to Lambton Castle, who had gone fishing while everyone else from the town of Washington close by attended church. Cursed for his blasphemy, Lambton sought redemption by journeying to the Holy Land on a pilgrimage after hurling his loathsome-looking catch into a nearby well. When he returned several

The Lambton worm, from an 1894 book of fairy tales.

years later, however, he was horrified to discover that the creature had grown into a monstrous worm that had been terrorising Washington's inhabitants by devouring their livestock, wilting their crops with its toxic breath, and killing anyone who had dared to challenge it. Even chopping it in two had been futile—the two halves simply rejoined straight away.

Happily, however, Lambton was able to rid the town of this accursed creature.

After seeking the advice of a wise witch, he commissioned the creation of an extraordinary suit of armour, bristling with long sharp spines. He then enticed the worm into the River Wear, and as soon as it attempted to crush him in its mighty coils, the spines on his armour sliced it into several segments, which were immediately carried away and dispersed by the river's fast-flowing current, so that they were unable to reconnect. Thus ended the Lambton worm's reign of terror.

This famous myth has been preserved through generations of retellings in northern England, but how it originated remains a mystery—unless, perhaps, the worm was in reality some less dramatic wild beast (or even an imaginative personification of a local now-forgotten natural disaster) that John Lambton had successfully combated? Having said that, however, eminent English historian and antiquarian Robert Surtees (1779-1834) claimed that during the early 19th Century he had seen at the village of Lambton (aka Old Lambton) in Washington a piece of preserved skin said to be from the Lambton worm, and which resembled the hide of a bull.

A similar ploy to Lambton's wearing of a spiny suit of armour was utilised in the killing of another British worm. This was when, during the 1300s, Sir William Wyvill wore a suit of armour covered in razor-sharp blades when battling an enormous worm at Slingsby in North Yorkshire. But this time, instead of a river dispersing the worm's body segments, the knight's faithful dog carried each segment away and buried it in a different location. Despite its resourcefulness, however, it could not save its master, or itself, from the worm's baneful effects. When it licked its grateful master's face, drops of the worm's deadly blood upon its jaws transferred inside its mouth and also onto the knight's face, and both man and dog died shortly afterwards.

This sad canine-induced fate features as a recurring motif in several other British worm tales too, including the killing of the Kellington worm by a shepherd and his dog, and the Loschy Hill worm's annihilation by Sir Peter Loschy and his dog.

The Linton worm was an immense worm that lived inside a cave in Roxburghshire on the Scottish Borders, emerging every dawn and dusk to prey upon livestock and their owners, and sporting a scaly hide so tough that no weapon could penetrate it. Fortunately, it was eventually dispatched by William de Somerville, Laird of Lariston, who used his horse to tempt the worm to open wide its enormous jaws—and, as soon as it did so, he rammed down its throat a long sharp lance upon whose tip was impaled a burning wheel bearing a huge lump of peat covered in boiling tar and brimstone. Not surprisingly, this deadly combination soon killed the worm, and the unusual topography of the region visible today is said to have been created by it during its writhing death throes. Alternatively, it could conceivably be that this topography is what inspired the legend of the Linton worm.

The impressions left by the coils of another massive worm are said to explain similar topographical features present upon Bignor Hill in West Sussex.

Not surprisingly, preserved relics from slaughtered dragons are not exactly commonplace, but after a huge worm with toxic breath was slain by a medieval knight at the hamlet of Sexhow in North Yorkshire, the villagers were so delighted that they skinned its carcase and hung its scaly hide inside the church at nearby Stokesley-in-Cleveland, where it remained on display for many years thereafter. Tragically for science in these enlightened modern times of DNA analysis, however, the skin is not there now, so we'll never know just what (if indeed anything?) was slain all those centuries ago at Sexhow.

THE BASILISK

Far smaller in size but no less deadly than the worms was the basilisk, one of the earliest references to which appeared in Pliny the Elder's magnum opus *Natural History* (c.77-79 AD). On first sight, this inconspicuous serpent dragon, just a few feet long, simply resembled a slender brown snake. On closer inspection, however, it could be seen to bear a regal crown of gold upon its head (hence 'basiliskos'—Greek for 'little king'). And when it moved, it raised much of its body vertically upwards in a proud, fearless stance, eschewing the lowly belly-crawling mode of locomotion typifying ordinary serpents.

The basilisk had every reason to be fearless. Its merest glance was enough to kill almost any living creature instantly—including another basilisk (and even itself if

it somehow caught sight of its own reflection). The tiniest drop of venom dribbling from its jaws was ample to poison the earth upon which it fell, or the water into which it dripped. And the faintest breath that it exhaled was sufficient to transform the land for many miles in every direction from fertile pastures into arid desert. Indeed, the very existence of the deserts where the basilisk lived, in North Africa and Arabia, was said to have been directly caused by its baleful presence.

In short, the basilisk was virtually invulnerable. Thankfully, however, it could be warded off with a sprig of the rue plant, and it could be killed outright by the rank odour of urine from a weasel. (This aspect of the basilisk legend may have been inspired at least in part by tales emanating from the Orient of cobras confronted and dispatched by mongooses.)

The basilisk's origin was just as uncanny as its appearance and capabilities, and also explained why this diminutive but much-dreaded monster was so rare. A basilisk was only created if the egg of a serpent were hatched by a rooster (cockerel)—a bizarre event which (thankfully!) was hardly likely to happen very frequently.

Despite its formidable nature, medieval alchemists were very keen to possess the ashes of a dead basilisk, because they believed that this rare, precious matter could transmute silver into gold. Some scholars, such as Theophilus Presbyter (fl. c.1070-1125), even believed that a basilisk could be magically created via a detailed recipe of ingredients and reactions, and that the resulting creature would convert copper into Spanish gold.

During this same period in history, moreover, the basilisk underwent a remarkable metamorphosis in popular legends and myths. Beginning with its acquisition of legs (usually just two, but sometimes several pairs), eventually it transformed entirely from its original serpentine form into a much more elaborate dragon of half-reptile half-bird form, known as a cockatrice, as documented later in this chapter.

Illustrations of a basilisk (right) and cockatrice (left), from Pietro Candido Decembrio's *The Animal Book*, 1460.

In bygone ages, the town of Baunei in the province of Ogliastra on the Italian island of Sardinia was terrorised by a basilisk-like serpent dragon that lived in the bushes there and was known as the scultone or ascultone. Just like the basilisk, its gaze was sufficient to kill anyone or anything that looked directly at it, but unlike the basilisk it was immortal. Nevertheless, Peter the Apostle finally managed to rid Baunei of this menace using a mirror, though the precise manner in which he accomplished the feat remains unclear.

In Gambia, Senegal, Guinea, and Guinea-Bissau in West Africa, traditional lore tells of a greatly-feared serpent dragon known as the ninki-nanka. Not only did it possess supernatural powers, it also concealed a precious diamond inside its head, from which it drew these powers. Moreover, echoing the odd manner in which a basilisk was created, a ninki-nanka was only hatched from an egg that was present in the very centre of a clutch of normal python eggs.

South America's basilisk equivalent, the basilisco, was toad-like rather than serpentine, and could only be hatched from a black egg. Accounts of this creature come from Santiago del Estero, the Mapuche area, and the northwestern region of Argentina.

THE NAGA AND OTHER ASIAN SERPENT DRAGONS

In Asia, some of the earliest serpent dragons on record were the nagas, which appeared in such ancient Indian works as the *Mahabharata*—a major Sanskrit epic whose oldest preserved text portions date back to around 400 BC, with its story's origins probably dating back to between the 8th and 9th Century BC. Nagas feature extensively in Hinduism, Buddhism, and associated religions, and were often venerated as nature deities. They were usually depicted with the bodies of enormous serpents, yet frequently possessed several heads, each of which bore a great cobra-like hood but sported the long jaws of a dragon. Sometimes, nagas could transform into humans, and on occasion they were even portrayed as human-headed serpents;

mostly, however, they assumed the form of hooded, many-headed serpent dragons.

One of the most famous naga images in Buddhism, reproduced in countless statues and paintings, is of the Lord Buddha (fl. 5th Century BC) sitting meditating in the lotus position upon the coils of the naga king Muchilinda, whose hooded heads, raised over his own, are protecting him from a rainstorm.

Human-headed naga statue in Bangkok's Royal Palaces, Thailand. (Dr. Karl Shuker)

Generally but not invariably benevolent, nagas were often linked to certain specific locations or topographical features, such as a stream or mountain. So too was the Indo-Malayan nogo, which was directly derived from India's nagas, and, like them, was worshipped as a deity by early Hindus in this region of Asia.

Greatest of all nagas was Ananta Shesha. This was a primal serpent dragon of creation, sporting a thousand heads whose hoods supported all of the planets in the entire universe, and who bore upon its immense back the Hindu god Vishnu when he was sleeping. When Ananta Shesha first uncoiled its infinite body, time began and the universe was created, but if it should ever coil up again, time would cease and the universe would come to an end. Fortunately, this great naga occupies itself in perpetually singing the glories of Vishnu in an unending paean of praise from each of its thousand mouths.

Another similar entity to the Indian naga, known as the neak, featured in Cambodia's Khmer lore. Those with an odd number of heads symbolised male energy, whereas those with an even number symbolised female energy.

In Korean mythology, the imoogi was a giant python-like serpent dragon living in caves or pools and usually benevolent, which after a thousand years transformed into a fully-formed Eastern dragon. Seeing one was good luck.

RAINBOW DRAGONS AND LIGHTNING DRAGONS

Rainbow dragons or rainbow serpents—enormous primeval serpent dragons inhabiting deep permanent water-holes—feature in the mythology and traditional beliefs of several widely-separated cultures, but most famously those of various African tribes and also in the Dreamtime lore of the Australian aboriginal nations.

According to many Australian aboriginal traditions, accurately preserved orally for 10,000 years, the world was created at least in part by the movements of gigantic polychromatic serpents that are seen today as rainbows. They are still worshipped as omnipotent nature entities that can bring good fortune but also spread disease, especially if angered.

The generic Australian aboriginal rainbow serpent occurs in native traditions spanning Australia, and is thus known by many different local names. These include: the mindi (in Melbourne, Victoria), andrenjinyi (Pennefather River, Queensland), neitee (Wilcannia, New South Wales), arkaroo (Flinders Range, South Australia), wanamangura (Laverton, Western Australia), and numereji (Kakadu, Northern Territory).

In the voodoo traditions of Benin (formerly Dahomey), a former French colony in West Africa, as well as in the Caribbean nation of Haiti, the rainbow dragon or serpent Ayida-Weddo was a spirit deity associated with fertility, rainbows, and snakes.

A very different African rainbow dragon was the mulilo of Zambia and the Democratic Republic of Congo (formerly Zaire). According to native lore, the mulilo only appeared if a rainbow rested upon a hillside, but this was its lone connection to a rainbow, because it certainly didn't resemble one. Instead, the mulilo, almost 6 ft long, and over 1 ft wide, looked like a loathsome coal-black slug-serpent, and like typical members of the worm category of serpent dragon its foul breath left disease and death in its wake.

The only way to kill one of these vile beasts was to bait a spike-lined cage with a living cockerel and wait until the bird no longer crowed. The cage could then be approached safely, as a mulilo would be found dead inside it, impaled by the spikes during its capture and killing of the hapless rooster. Portions of this rainbow dragon's black, shrivelled flesh were supposedly worn as fertility charms by local hunters, though whether these objects were truly from a mulilo is another matter entirely.

In Australian aboriginal lore, lightning dragons were enormous serpent dragons that normally dwelt exclusively in the sky, but jumped down to Earth during storms and then leapt back up into the sky, the energy that they released during these exertions creating lightning. Similar entities also occur in traditional Zambian folklore, ensheathed in thick crocodilian scales and sporting long goat-like horns upon their heads. In the legends of the Native American nation of western Vancouver, Canada, the supernatural entity He'-e-tlik became life-long friends with Tootooch, the thunder bird, after which he transformed into the lightning snake and accompanied Tootooch everywhere, even killing whales for him to devour by shocking them to death via contact with his electricity-generating serpentine body.

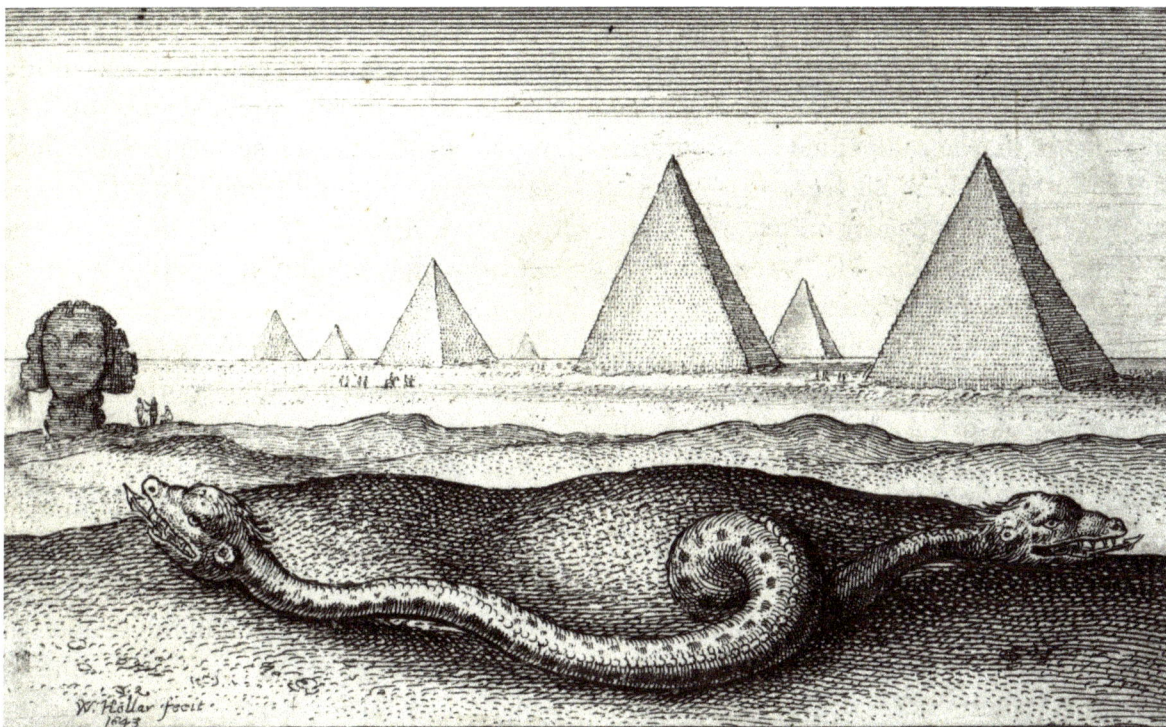

Amphisbaena, by Wenceslaus Hollar, 1643.

THE AMPHISBAENA

A serpent dragon no bigger but even weirder than the mulilo was the amphisbaena, on account of its two heads—or, to be specific, the sites of these heads. For whereas there are many legends of dragons with more than one head, their heads were always present at the front end of their body. Uniquely, however, as documented as long ago as the 2nd Century BC by Greek poet Nicander of Colophon, the amphisbaena had a head at each end of its body, and could therefore move in either direction—sometimes accomplished by grasping one head in the jaws of the other so that its body became a hoop that could roll rapidly over the ground.

An amphisbaena was almost impossible to approach unseen, because only one head slept at a time, the other one staying awake, particularly when this creature was laying eggs. And just like many serpent dragons of the worm category, if an amphisbaena were cut in half, the two segments would promptly rejoin.

According to Greek mythology, the amphisbaena was spontaneously generated from drops of blood falling onto the desert sands from the severed head of the gorgon Medusa when her slayer, the hero Perseus, flew over Libya with it on his journey back home to the Greek island of Seriphos. Although the amphisbaena's principal diet was ants, it was claimed by some writers to be extremely venomous, and one was blamed for the subsequent death of Mopsus, a seer who was also one of the famed Argonauts that accompanied Jason on his quest for the Golden Fleece.

Yet despite its deadly nature, the dual-headed amphisbaena was often cited by early scholars for its medicinal qualities. Sometimes a living specimen was needed, otherwise the skin of one was sufficient. Among the assorted ailments that it reputedly eased were arthritis, chilblains, and the common cold, as well as assuring a safe pregnancy, and keeping warm during the winter if working outside. Eating the meat of an amphisbaena could even attract lovers, and killing one during a full moon would imbue its slayer with great power provided that he was pure of heart and mind.

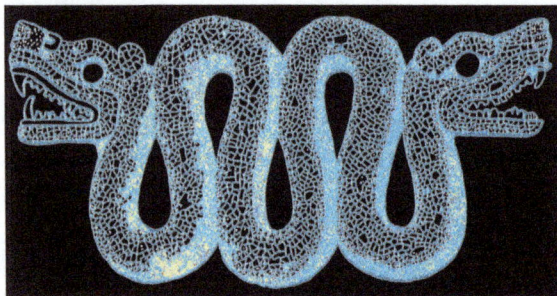

Xiuhcoatl, represented as a vertical turquoise amphisbaena. (Dr. Karl Shuker)

Amphisbaenas often featured in Mesoamerican and Inca cultures too, frequently depicted with a vertically undulating body, and symbolising eternity. Some of the most spectacular renditions are composed of turquoise mosaic, a stone believed by the Aztecs to emit smoke, and therefore a very fitting mineral for portraying a dragon, especially in versions representing Xiuhcoatl—known variously as the fire serpent or the turquoise serpent. In these New World versions, one head was sometimes much larger than the other, rather than always being identical as in the original Old World amphisbaena.

In Chile, the oral traditions of the Elqui villagers tell of a 6-ft-long spotted amphisbaena known as the culebrón. During the day, it crawled very slowly upon the ground, but at night it took flight, because, uniquely among amphisbaenas, this version sported a pair of wings.

Perhaps the strangest South American amphisbaena, however, was the manora, whose basic form resembled a giant earthworm. Its head and tail ends were indistinguishable from one another, but its body was covered all over with sharp feather-like quills.

Today, the legendary amphisbaena gives its name to a group of real-life reptiles, the amphisbaenians, which are also known as worm lizards. Their heads are so similar in appearance to their tails that it can be difficult to distinguish which end is which, thus recalling the two-headed amphisbaena.

Having said that, just like the basilisk the amphisbaena underwent a profound transformation during medieval times. It gained not only a pair of legs but also a pair of wings, as well as a well-delineated tail—at the end of which was its second head. It also acquired the literally petrifying, gorgonesque ability to turn anyone who

19th-Century engraving of an amphisbaenian.

some of the marine examples included the largest dragons ever known. Chief among these was Jormungander, the fearsome Midgard serpent of Norse mythology, one of the foul offspring of Loki, god of evil. A worm with foetid breath and limitless coils, once its existence was discovered by the gods Jormungander was hurled into the vast ocean by the thunder god Thor. Here it was destined to remain through all the ages that followed, encircling the whole world with its head permanently grasping its tail (an ancient dragon image referred to as an ouroboros, signifying eternity, immortality, and rebirth) until Ragnarok, the Twilight

looked at it to stone with just a single glance. This advanced version of the amphisbaena is known as the amphisien, and commonly occurs in heraldry.

MARINE SERPENT DRAGONS

Some of the world's most famous serpent dragons were of the aquatic variety, and

Jormungander the Midgard serpent battling Thor during Ragnarok, by German Art Nouveau illustrator Emil Doepler, 1855-1922.

of the Gods. Only then would Jormungander finally release its tail and confront Thor in a mighty one-to-one battle in which both would perish.

Prior to that apocalyptic event, however, Thor encountered this monster on two separate occasions. One of these was when he inadvertently hooked it while fishing with a companion in a boat on the sea. When Thor's companion saw the colossal, hideous head of Jormungander appear above the water surface, however, he was so terrified that their boat would be capsized by the great sea dragon that he cut Thor's line, releasing Jormungander to sink back beneath the waves, and leaving an enraged thunder god to roar impotently in its wake.

The other occasion was when Utgarda-Loki, king of the giants, deceived Thor with the magic of disguise into believing that Jormungander was a mere house cat, and dared him as a test of his strength to lift it off the floor—clearly an impossible task, but which, to the giant king's horror, Thor almost achieved!

The earliest major sources for myths concerning the Midgard serpent all date from at least as long ago as the 13th Century. They are: the *Prose Edda* (a four-volume Icelandic work recording the Norse myths and examining verse forms used in old Norse poetry); the skaldic poem *Húsdrápa* (describing Thor's fishing encounter with Jormungander); and the Eddic poems *Hymiskviða* (also documenting his fishing encounter) and *Völuspá* (describing Thor's fatal confrontation with Jormungander during Ragnarok).

Britain's answer to the Midgard serpent (and, indeed, derived from the early Norse myths

Jormungander hooked by Thor, from a 17th-Century Icelandic manuscript.

concerning the latter monster following Norway's annexing of the Orkney Islands in 875 AD and their settlement by the Norse) was a marine mega-dragon called the meister stoor worm. Just like Jormungander, this mighty serpent dragon encircled the entire globe in its coils, causing the sea to surge and the earth to shudder if it so much as opened its mouth to yawn. Like all worms, its breath was toxic, and its huge tongue was powerful enough to sweep entire villages into the sea with a single casual flick. After eventually settling off the coast of northernmost Scotland, this monstrous marine worm was kept satiated by being fed seven young virgins each week, until finally it was time for the king's own daughter to be sacrificed.

In desperation, the king offered her hand in marriage to anyone who could save her from the stoor worm's maw, and a brave young man named Assipattle rose to the challenge, putting into operation a very dramatic variation of the ploy used so successfully by the Laird of Lariston with the Linton worm. Setting out in a little boat that transported him and an iron pot filled with burning peat, in accordance with his plan Assipattle was duly carried by the waves deep inside the stoor worm when it opened its vast mouth. Eventually, the ingested waves carried his boat to the worm's liver, and here Assipattle lost no time in slicing a gaping hole inside this huge vital organ and filling it with the peat from his pot.

As its liver began to smoulder, the great worm retched in shock and pain, vomiting

Assipattle out of its mouth, after which the dying monster thrashed in agony before exploding. Parts of its body became islands when they fell back into the sea, including the Orkneys, the Shetlands, and Iceland, with its perpetually-aflame liver transforming into a series of volcanoes on Iceland.

A comparable marine serpent dragon of Scottish provenance was the grey-crested cirein croin. This monster was of such immense proportions that it could swallow an entire whale in a single gulp.

According to Celtic legend, the Irish seas were home to a huge serpent dragon too. Known as the sinach or muirdris, it sported a horned head, glaucous scales, countless body spikes, and could swell up like a balloon.

Featuring in Icelandic folklore, the skirimsl was another immense northern serpent dragon of the sea, and had originally been a ferocious man-eater, but was eventually subdued, bound, and thus rendered harmless by Bishop Gudmund—or at least until Doomsday. Then, just like the Midgard serpent, it will escape its shackles—leaving chaos and destruction in its vast watery wake.

Yet another gargantuan serpent dragon of the sea was the biblical Leviathan. Immensely long, ensheathed in glittering mail, with over 300 incandescent eyes, this stupendous beast emitted blasts of red-hot steam from its nostrils and fire from its enormous jaws. Shaped and breathed into life by God on the fifth day of Creation, Leviathan was so immense that it devoured dragons measuring over 1000 miles long as if they were mice. God initially created a pair of these spectacular sea monsters, but, fearful that if they established a thriving lineage of their kind they would annihilate the world, He later destroyed one of them, but to compensate the other for its loss, He granted it immortality.

Several Greek myths refer to monstrous serpent dragons of the sea. During the Trojan War, Laocoön, a priest of Poseidon, voiced his suspicion that the wooden horse of Troy given by the Greeks was some sort of trick, not to be trusted, and begged for it to be destroyed. In response, the Greeks' divine supporter, the goddess Athena, sent two enormous sea dragons with crests of fire through the waters until they reached Laocoön, whereupon they emerged and strangled him, as well as his two sons.

And it was from a gigantic marine serpent dragon named Cetus, limbless but sporting fins upon its head

'Destruction of Leviathan', by Gustave Doré, 1865.

A representation of Cetus. (Tim Morris)

and sinuous body, that the hero Perseus successfully rescued the princess Andromeda. She had been chained to a cliff as a sacrifice to Poseidon by her father Cepheus, king of Ethiopia, in order to appease the sea deity after he had been angered by Cepheus's boastful wife, Cassiopeia—who had rashly bragged that she was more beautiful than Poseidon's daughters, the nereids or sea nymphs.

In Aztec mythology, the cipactli was a marine serpent dragon of leviathanesque proportions that resembled a finned crocodile. Out of its colossal carcase the Aztec deities created our entire planet.

FRESHWATER SERPENT DRAGONS

Smaller but otherwise similar aquatic serpent dragons to many of the giant marine versions also existed in lakes and rivers. For instance, various freshwater counterparts to Iceland's monstrous sea-dwelling skirimsl were said to exist in this country's rivers, especially the Lagarfljót. Its representative was called the lagarfljótsormurinn, and is occasionally reported even today.

A singularly spectacular freshwater serpent dragon, blue in colour with golden crests and sparkling eyes but venomous breath and a stinging tail, guarded the Castalian Spring in Boeotia, central Greece. However, it was slain by Cadmus, a Phoenician prince who then founded the city of Thebes there.

Equally noteworthy was medieval France's gargouille, which emerged one day in 520 AD from the River Seine at Rouen, capital of Normandy. For whereas most dragons emitted fire or toxic vapour from their jaws, this remarkable aquatic dragon spouted forth great torrents of water, flooding the surrounding countryside, and washing away entire villages, their inhabitants drowned or devoured by it. To end the gargouille's tyranny, St. Romain, archbishop of Rouen, boldly confronted it in its great cave, but before this huge water worm had chance to open its dread jaws and engulf him in a stream of chilling fluid, the saint raised his hands before it and held his fingers in the shape of the Cross.

Instantly, the gargouille's seemingly unquenchable reserves of water dried up, and this once untameable dragon was transformed into so passive a creature that it even permitted St. Romain to bind its neck and lead it back into Rouen. Here, however, the survivors of its erstwhile reign of terror were not so forgiving, and the gargouille was duly burnt to death. Yet even

A dragon-shaped gargoyle.
(Pixabay/Arjane)

today it still lives on, insomuch as its name has been given to those famous carvings of grotesque monsters on churches that spout forth rainwater—gargoyles.

This famous legend dates back at least as far as the early 1200s, each year since when a magnificent pageant would be held

Cadmus and the dragon.

on Ascension Day in Rouen to commemorate the city's salvation from the dreaded gargouille. Part of this pageant would feature a marching parade in which a beadle resplendently adorned in violet livery would bear aloft on a pole a wicker effigy of this water-spouting monster, in whose mouth a live animal such as a rabbit or fox cub or even a suckling pig would be held. This annual celebration continued until the French Revolution.

Ireland's lakes were once home to very large freshwater serpent dragons known as peistes or piasts, measuring up to 20 ft long, and sporting a thick mane of hair down the centre of their back. When St. Patrick banished all snakes from the Emerald Isle, however, he also exiled the peistes—except for a particularly aggressive venom-fanged individual with ebony scales the size of dinner plates and ram horns curled around its ox-like ears. Fortunately, this peiste was eventually bound securely at the bottom of Lough Foyle by St. Murrough.

The biggest peiste of all was the green-scaled ollipeiste, which had originally been a very inoffensive dragon despite its great size. However, it became so enraged when it discovered what St. Patrick was doing to all its brethren that its furiously-thrashing tail carved out the Shannon Valley before it too was forced to leave Ireland, making a new home for itself in the sea.

North America's lakes and rivers are associated with numerous legends and folk stories of aquatic horned serpents but also some concerning bona fide serpent dragons, including one that resembled a huge horned leech and was known as a weewilmekq. Huge swamp-dwelling serpent dragons remain an intrinsic component of native lore for many Indian tribes inhabiting the rainforests of South America too. One in particular was the camoodi of Guyana, whose body looked very much like that of a colossal anaconda, but whose head was a dragon's, bearing a pair of horns. Similar entities also feature in the folklore of the Tarahumare Indians living in northwestern Mexico's Sierra Madre range.

As for Australia, back in the far-distant Dream-Time a noticeably deep water-hole present at the junction of the Wollondilly and Wingeecaribee Rivers in New South Wales was supposedly frequented by a huge serpent dragon known as the gurungatch. It resembled a lizard in overall form but possessed only fins instead of limbs, was brilliantly adorned in scintillating scales of green, gold, and purple, and had eyes like shining stars. The gurungatch enjoyed preying upon humans, but despite being relentlessly pursued by the great fisherman Mirragan, it always evaded capture. Finally, however, it was confined within a second water-hole, and may still be there today!

TWO-LIMBED WINGLESS DRAGONS—THE LINDORMS

After the serpent dragons, the next stage upon the dragon's evolutionary ascent was the lindorm. This was an elongate wingless semi-dragon usually up to 20 ft long (but sometimes even bigger) that possessed a single pair of limbs. (In Sweden, however,

A Swedish lindorm, although technically this particular specimen is a serpent dragon because it is limbless. (Richard Svensson)

the term 'lindorm' is also applied to serpent dragons.)

In cases where these limbs were fore-legs, the lindorm when resting maintained its serpentine body in a vertical series of coils, thereby raising its head and forequarters high above the ground. In cases where they were hind legs, the lindorm walked in a bipedal manner, rather like theropod dinosaurs such as *Tyrannosaurus* and *Allosaurus*. Indeed, so successful was this bipedal mode of locomotion that lindorms with hind legs could supposedly out-run a man riding on horseback.

Lindorms occur extensively in the folklore of Scandinavia (especially Sweden, where some lindorms sport a small pair of wings instead of limbs) and central Europe (notably Switzerland), inhabiting mountainous localities, as well as churchyards—where they devoured human corpses.

Sometimes, they would even invade churches.

As with many serpent dragons, lindorms possessed a venomous bite rather than breathing fire like classical dragons. In addition, it was claimed that anyone finding a shed lindorm skin would gain great knowledge in the fields of medicine and natural history.

A Swiss lindorm, from *Itinera per Helvetiae Alpinas Regiones Facta Annis 1702-1711* by Johann Jakob Scheuchzer, 1723.

THE HYDRA AND OTHER VARIATIONS UPON THE LINDORM THEME

As with all dragon categories, there were a number of very unusual, atypical lindorm varieties. These include the following examples.

As documented in the Germanic Nibelungenlied, one of the most famous lindorms was the ferocious version into which the dwarf Fafnir metamorphosed as a result of the lust and greed for the treasure of his father (a miserly dwarf king) that had corrupted his soul after he had killed his father in order to seize the treasure for himself.

The knucker, as depicted in Andy Paciorek's book *Strange Lands*. (Andy Paciorek)

Another well-known example of a transformed lindorm, this time from Scandinavian mythology, was Prince Lindorm. Although human, due to some mismanaged magic he was born a lindorm instead, but eventually he acquired his real form by gaining the true love of a willing human bride.

Perhaps the most familiar British lindorm was the knucker of Lyminster in West Sussex, England. This particular lindorm was amphibious, terrorising the local populace on land, but living in a deep pool known as the Knucker Hole near Lyminster's church. Eventually, however, its unwelcome activities came to an end when it was surreptitiously killed with a huge poisoned pudding. According to local lore, knuckers and knucker holes once existed in several other Sussex locations too.

Less well known was the European aspis. This was a relatively small lindorm, but lethal nonetheless, because its scaly skin was poisonous to the touch and its bite resulted in immediate death. Luckily, it was rendered docile if exposed to mellifluous strains of music.

The hydra as portrayed in Conrad Gesner's *Nomenclator Aquatilium Animantium*, 1560.

By far the most unusual, and deadly, dragon of basic lindorm form, however, was the Lernean hydra—whose slaying constituted one of the twelve great labours of the Greek hero Heracles (Hercules in Roman mythology).

Although this monster is usually depicted as wingless and only two-legged, it was more than ably compensated by virtue of its numerous heads (generally given as nine, but sometimes only seven), each borne upon a separate neck. And each time that a head was cut off, two new ones grew in its stead, until Heracles successfully countered this by burning each neck as soon as its head was lopped off.

TIAMAT THE BABYLONIAN SEA LINDORM

No less terrifying than the hydra was the monstrous lindorm that the primordial Babylonian sea goddess Tiamat became in order to take vengeance upon her numerous divine offspring for murdering their father, Apsu—who was Tiamat's husband. Her immensely long serpentine body was plated with countless impervious scales, her single pair of powerful muscular forelimbs were armed with razor-sharp talons, and upon her head, borne proudly by her tall erect neck, stood forth a sturdy pair of curved horns. In some versions of this myth, Tiamat had several heads and necks, just like the Lernean hydra. Happily, however, for the new generation of deities and humanity, Tiamat was ultimately slain by Marduk, the sun god.

The Tiamat legend occurs as far back as the *Enûma Eliš*. This is the Babylonian creation epic, which was recovered in fragmentary form from seven clay tablets collectively bearing approximately a thousand lines of Akkadian cuneiform text and dating from the 7th Century BC (though the original composition of the text may date back at least a further four centuries). The tablets were discovered in the ruined Royal Library of Ashurbanipal (the last great king of the Neo-Assyrian Empire) at Ninevah (located today in Mosul, Iraq) by Austen Henry Layard in 1849, and their text was first published by George Smith in 1876.

The basic storyline or motif of a deity or hero slaying a primordial (and sometimes multi-headed) dragon representing chaos as typified by the Tiamat legend is known as a chaoskampf, and occurs in a number of other early Middle Eastern myths too. Although in these, the dragon is usually a straightforward serpent dragon—as indeed was Tiamat too in certain versions of her legend—rather than a lindorm.

For instance, Lotan, an immense seven-headed dragon of chaos in ancient Mesopotamia that encircled the world like the Midgard serpent of Norse mythology, was triumphantly dispatched by the Canaanite deity Baal following a stupendous battle. Similarly, Mushmahhu was an enormous seven-headed anguinine dragon slain by the Sumerian deity Ningirsu. And Teshub (=Tarhunt), the weather god of the ancient Hittites, slew Illuyanka, a colossal seven-headed serpent dragon of chaos, before strewing its blood over the earth to fertilise the crops.

Even in India, there is an early Vedic legend of how Indra the Hindu weather god slew Vritra—a gargantuan world-encircling serpent dragon with three heads. Vritra was itself of divine status, but refused to release much-needed rain upon the parched earth during periods of drought until defeated by Indra.

TWO-LIMBED WINGED DRAGONS—THE WYVERNS

In evolutionary terms, one stage up from the lindorm was the wyvern (a name derived from the French for 'viper'), which began to be depicted in European carvings around the early 11th Century AD. This European semi-dragon generally had the basic form of a hind-limbed lindorm (more rarely with forelimbs instead) but with the addition of a single pair of large bat-like wings, and usually a venomous scorpion-like sting at the tip of its long coiled tail.

A typical wyvern, holding a fleur-de-lis.

The wyvern frequently appears as a heraldic beast, and originally personified loyalty and bravery, but once its image began to be perceived as evil, it became a heraldic representation of war and plague instead. There is also a marine heraldic version in which the wyvern's sting-bearing tail is replaced by a fishtail.

Like lindorms and serpent dragons, the breath of the wyvern was toxic, spreading pestilence, corruption, and death. And although it preyed upon livestock and other animals, the wyvern had a particular taste for human flesh—especially that of young maidens. A few wyverns evolved the ability to breathe fire, thus constituting the lowest rung on the dragon's evolutionary ladder to have achieved this feat.

Some of the most notable wyverns feature in British folktales and legends. One of these was of the fire-breathing variety, which inhabited the abandoned ruins of the castle formerly standing in Newcastle Emlyn, in Dyfed, Wales. This dread beast effortlessly resisted all attempts to kill it, because its scaly skin was invulnerable to weapons—except, that is, for a single small, unprotected spot. Just like bulls (at least according to folklore), some dragons were enraged by the mere sight of the colour red, so one wily soldier threw a large piece of red flannel into the nearby river—and when the wyvern emerged from its hideout to chase after it, the soldier lost no time in killing the creature by shooting it in its weak spot. Perhaps uniquely among all dragons, this wyvern possessed a navel!

In Mordiford, Herefordshire, a little girl called Maud successfully reared a baby fire-breathing wyvern, as green as a cucumber and not much bigger than one either, which she had found in a wood, lost and lonely. When, however, it grew to full size as an adult and became highly aggressive towards people (though never to her), eventually slaughtering villagers for food, it was duly dispatched by a valiant knight.

An unusual freshwater wyvern with webbed feet once dwelt in Cynwch Lake neat Dolgellau in Clwyd, Wales, and was sometimes seen lying upon the lake's shore and even on the slopes of a nearby mountain, Moel Offrum, moving forward in a series of looping movements like a gigantic caterpillar but leaving an acidic trail behind it resembling that of a slug. Once it began attacking the locals, however, and contaminating the countryside with its noxious breath, several attempts were made to kill this amphibious bane. Eventually, a shepherd boy named Meredydd crept up behind it one day while it was out of the water, sleeping on a hill, and chopped off its head with a magic axe.

The wyvern of Brent Pelham in Hertfordshire was notable for sporting a beard and a pair of large ears alongside its more usual features.

THE JACULUS AND ZILANT—
WYVERNS WITH A DIFFERENCE
In addition to the typical wyvern form, there were certain other versions that displayed some interesting variations upon this particular morphological theme. Notable among these were the jaculus and the zilant.

Referred to as long ago as the 1st Century AD by the Roman poet Lucan (39-65 AD) in his epic multi-volume poem *Pharsalia*, the jaculus was said to resemble a 10-ft-long winged serpent but with a pair of short forelimbs too, so it presumably resembled an unusually elongate wyvern. It also possessed two tongues, one of which

was barbed, and had a passion for collecting gemstones and gold.

When not doing so, however, the jaculus spent much of its time concealed from sight high in the Arabian spice trees that it jealously guarded—hurling itself ferociously like an animated spear (hence its name—'jaculus' translates as 'javelin') at anyone approaching below, then biting them deeply and lethally in the neck or throat with its long venomous fangs.

The zilant appears in Russian and Tatar mythology and resembled a typical wyvern in basic form except for lacking a sting at the tip of its tail and sometimes sporting two heads instead of just one. It occurs in several legends associated with the founding of Kazan, the capital city of Russia's Tatarstan Republic, and features in Kazan's coat of arms. It was first made this city's official symbol in 1730. According to traditional lore, any snake that lived over a century became a zilant.

THE COCKATRICE
During the Middle Ages, the small yet deadly North African serpent dragon known as the basilisk underwent a very dramatic transformation in mythology, metamorphosing into a much bigger, truly grotesque type of wyvern-like dragon known as a cockatrice. It gained a pair of large bat-like wings, a long coiled tail (still covered in scales but often terminating in a sharp sagittal tip), and a single pair of sturdy rooster-like legs that enabled it to walk upright. Furthering its cockerel parallels, however, it also sported a coxcomb on its

Zilant statue in Kazan. (Snowleopard/Wikipedia)

head, a pair of pendulous facial wattles, a pointed horny beak, sometimes a covering of feathers upon its body, and even the ability to crow like a farmyard rooster too.

Even so, this weird reptilian bird (or avian reptile?) retained the basilisk's deadly gaze, its dread of weasels and rue, and also, albeit in a reversed version, its bizarre mode of creation. Now, one of these monstrous entities would arise if a round leathery shell-less egg laid by a seven-year-old cockerel when the dog star Sirius was in the ascendant was hatched by a toad in a dung heap. Although such an occurrence may seem highly unlikely, cockatrices were reported not just in North Africa like their basilisk antecedent but also widely through Europe, including several examples from Britain.

One of the most recent of these was the Renwick cockatrice. Crowing loudly, this huge bat-winged horror, black in colour but sporting facial wattles and a coxcomb, emerged from the foundations of a church being demolished by workmen in the Cumbrian village of Renwick in 1733. Happily, the monster was dispatched by local hero John Tallantine with a sturdy wooden lance hewn from a rowan tree—famed for its magical, evil-repelling properties.

Iceland is not a country well known for dragon legends, but its traditional lore does lay claim to its own version of the cockatrice—a deadly creature called the skoffin. It has a very curious origin—the highly unlikely outcome of a liaison between a male fox and a female cat (the offspring of the reverse pairing—between a tom cat and a vixen—is called a skuggabaldur, and is just as ferocious as a skoffin). Similar in form to the cockatrice, the skoffin also shared the latter's lethal gaze—and could even kill an-

Continuing this intriguing link between cockatrices and Christianity, it may come as a surprise to learn that the cockatrice was formerly mentioned no less than four times within the Old Testament of the Bible, three of these mentions occurring in the Book of Isaiah, and the fourth in the Book of Jeremiah. The Hebrew terms that were once translated as 'cockatrice' were 'Tsepha' and 'Tsiphoni', but in modern-day versions they are translated as 'viper' or 'adder' instead, thus explaining the cockatrice's disappearance from this holy book.

Iceland is not unique in boasting its own specific form of cockatrice. So too does Korea—the gye-ryong ('chicken-dragon'). Just as Oriental dragons are generally more benevolent than malevolent, however, so too is this Korean cockatrice, often pulling the chariots of notable legendary heroes or those of their parents.

Cockatrice under attack from a rue-bearing weasel, by 17th-Century Bohemian etcher Wenceslaus Hollar.

other skoffin simply by looking it in the eye. Otherwise, it was virtually invincible, unless shot with a silver button upon which the sign of the Cross had been inscribed.

FROM THE WINGED two-limbed wyvern, the next level in the dragon's evolutionary advancement was the development of a second pair of limbs—yielding the quadrupedal classical dragon.

CHAPTER 2:
THE NATURAL HISTORY OF HIGHER DRAGONS

HIGHER, COMPLEX, or advanced dragons are the ones that most readily come to mind when people think of dragons—huge, burly, fire-breathing monsters with four (or more) powerful limbs and sometimes a pair of mighty bat-wings too in the West, or ethereal, sky-dwelling, quadrupedal deities in the East. This chapter surveys the world's dazzling diversity of higher dragons.

FOUR-LIMBED WESTERN DRAGONS WITH/WITHOUT WINGS—THE CLASSICAL DRAGONS

Classical dragons constitute the most familiar category of Western dragon—the true dragon for many people—and existed in two distinct forms, winged and wingless. Widely reported throughout Europe, North Africa, the Middle East, and even, more rarely, in the Americas, with counterparts in Oceania too, both versions were usually (though not invariably) malign in nature and antagonistic to humans. In general form, they were extremely large, four-legged reptiles, armed with fearsome claws on their feet and a long sweeping tail, plus a sturdy neck, a thick scale-armoured hide,

and often a serrated ridge of spines running down the centre of their neck and back. In addition, some such dragons were equipped with a huge pair of bat-like wings, but often reinforced with scales, and with spines at the ends of their pinions.

Winged or wingless, classical dragons that exhaled fire were termed fire-drakes. If they did not exhale fire, they were called cold-drakes.

A typical winged classical dragon.
(Richard Svensson)

Engraving of Winkelriedt battling the Mount Pilatus dragonet.

Winged classical dragons were often associated with hoards of treasure that they jealously guarded in high mountainous zones, when not soaring through the skies like dark monstrous shadows, spraying hapless towns and villages below with lethal blasts of flame.

Notable among these was the alklha or alicha. This was a gigantic primordial black dragon featuring in the mythology of Siberia's Buryat people. Its wings were so enormous that when it spread them forth across the sky they eclipsed the sun, and its massive talons dug out the craters on the moon.

Smaller winged classical dragons were called dragonets, such as Switzerland's Mount Pilatus dragonet. This diminutive yet nonetheless deadly beast was slain by a convict named Winckelriedt, convicted for manslaughter but released on the condition that he faced the dragonet. However, he also died when the creature's virulently toxic blood ran down his sword onto his bare hand.

In contrast, the wingless form of classical dragon was earthbound. It could often be found lurking in swamps or forests, when not scourging the countryside and holding kingdoms to ransom with dread demands of voluptuous female virgins to dine upon until eventually slain by some bold knight errant. Some lacked the fire-breathing ability of their winged brethren, exhaling noxious vapour instead, in the manner of serpent dragons.

Wingless four-limbed dragons of the classical dragon variety are technically known as lindworms. However, this term is often applied interchangeably with 'lindorm' for wingless two-limbed dragons too (as in heraldry), thus creating much nomenclatural confusion. For the sake of clarity, therefore, it has not been utilised in this book.

EUROPEAN CLASSICAL DRAGONS

Europe's myths and folklore contain an unparalleled profusion of winged and wingless classical dragons, of which the following is only a very brief overview.

Britain's dragons were predominantly of the limbless, worm type, but a few classical dragons have also been reported, particularly from Arthurian legends. Perhaps the best known example is contained in *The Mabinogion* (a 12th-Century collection of Welsh legends). It features two warring nations personified as dragons, plus the magician Merlin, when still a child during the early 5th Century AD—a time when parts of Wales were supposedly ruled by Vortigern, a semi-mythical warlord-king.

Vortigern had instructed workers to erect a tower at Dinas Emrys in Snowdonia, but every morning they saw that the tower's foundations had been mysteriously destroyed during the previous night, and its building materials had vanished. Angry yet perplexed, Vortigern consulted his court magicians, who advised him that the tower would only be completed if the blood of a boy born of a virgin was sprinkled over the site. Needless to say, such boys were not readily discovered, but at last Vortigern's minions found one—an uncanny boy-sage called Merlin.

According to legend, Merlin's father was not a man but a demon, and so, technically, he did not have a father (at least not a human one). Happily for Merlin, however, his powers forewarned him of his impending doom, and so he persuaded Vortigern to allow him to investigate for himself the mystery surrounding the tower's unstable foundations. Merlin soon discovered the real explanation—beneath the foundations was a subterranean lake in which two entrapped dragons, one red (symbolising Wales and its Celtic people) and one white (symbolising the Saxons), fought violently

The Welsh red dragon.

each night. Consequently, the dragons were freed, and flew away, still fighting, until the red dragon eventually won, and so became in time to come the national emblem of Wales. Merlin, meanwhile, was saved, and Vortigern could now build his tower untroubled by further terrestrial perturbations!

Another noteworthy British classical dragon was a savage fire-breather that had been terrorising the vicinity of Carhampton on the Somerset coast until tamed by the Celtic saint Carantoc. In return, King Arthur promised to find a lost altar that had served the saint as a boat when journeying by sea from Wales to Somerset.

One of the most famous Scandinavian examples of a classical dragon was an adversary of Beowulf. According to the Nowell Codex (the single surviving manuscript of this epic medieval poem), it was winged, fire-breathing, and treasure-guarding. Beowulf slew this dragon 50 years after defeating the monstrous Grendel and his equally formidable mother, but was mortally wounded in the fight.

Even more fearsome, and potentially catastrophic in its intention, however, was the nidhogg of Norse mythology. This foul beast lived at the base of Yggdrasill, the World Tree that supported the entire universe in its branches, and dedicated its loathsome existence to gnawing incessantly at one of this colossal ash tree's three great roots in the hope of destroying it, which in turn would herald the end of all things. Happily, however, the root was restored each day with water brought by the Norns or Fates from an enchanted well.

Sweden's dragons were avid treasure hoarders but remarkably diverse in form, as kindly elucidated here by Swedish fantasy artist Richard Svensson in an email to me of 4 March 2013:

The Swedish dragon is probably a bit different from many of its ilk in other cultures. Dragons in Swedish folklore are supernatural creatures,

A cat-headed, fish-bodied classical
dragon from Blekinge, Sweden.
(Richard Svensson)

and are said to be the souls of old misers, who keep money from their family. When such persons die and hide treasure that should rightfully become the inheritance of their family, their souls are transformed into a reptilian monstrosity that most often resembles a serpent. The dragon can be both huge and small. In my region [Blekinge, one of Sweden's southernmost provinces], it's most often quite small. Further up the country it can assume colossal proportions, and is

often seen flying through the night sky. It is described as being wingless and often surrounded by a cloud of sparks. There is also another interesting description that we'll get to in a moment.

Swedish dragons are seldom said to breathe fire, but they surround themselves with this element in various ways. They do spit venom and have hypnotic powers. Sometimes they can shape-shift into normal human beings or ordinary animals.

During the day the dragon resides in the place where the treasure is hidden, for example under a rock. The dragon will bore a funnel through the rock, which allows it to enter and exit its hiding place. Such funnels are called 'dragon pipes', and are probably based on some geological anomaly I'm not really aware of.

One really interesting detail about Swedish dragons is that they're often said to resemble a black rod flying through the sky with a burst of flame behind it. Basically a rocket, or a cigar shaped UFO.

Swedish dragons are invulnerable to any kind of man-made weapon. You kill it by having someone, preferably a priest, read from the Bible aloud, making it dizzy and crash to the ground. You can also shoot it with a lead bullet as long as you also load your rifle barrel with a rolled-up Bible page.

The dragons of Blekinge are generally described as having fish-like bodies, four legs, and the head of a cat. They're about the size of a grown man's leg, and are often found crouched over pots of money or graves containing hidden treasure.

Some Blekinge dragons are reported to look like large snakes, and are sometimes confused with the Swedish lindorm, which I might've told you about. However, the dragon will in its second year of being grow a pair of wings, turning it into one of those winged serpents that seem to pop up in several cultures.

A highly distinctive winged classical dragon from southern France was the drac. This very large river-dwelling reptile was able to become invisible at will, and could also shape-shift into a human. In the year 1250, it reputedly kidnapped a lavender-seller, imprisoning her in its underwater home for 7 years to rear its son, and was a notoriously wily opponent when confronted by errant knights and courageous villagers, killing over 3000 of them with impunity in the vicinity of Beaucaire before apparently dying of old age. Each year, the drac legend is celebrated in Beaucaire by a parade during 20-22 June.

Dragon-wise, Switzerland is best known for its lindorms, but other types have also been reported here. One was the elbst—a wingless classical dragon of aquatic lifestyle, with a huge porcine head, a long slender body, four clawed limbs, and a powerful tail. It inhabited Lake Selisbergsee near Lucerne, but at night it would come ashore to prey upon any sheep close by

Much more bizarre was a winged classical dragon that had boar-like bristles covering its back, and the face of a cat. This Swiss horror was allegedly encountered by a villager from Altsax called Andreas

An engraving from 1708 of Wangserberger's cat-faced classical dragon.

Roduner and a colleague while crossing the mountain of Wangserberger in or around 1660. When it saw the two men, it rose up onto its hind legs, but they managed to escape its clutches.

Speaking of cats: according to the traditional folklore of Lithuania and the other Baltic States, this is what a household dragonet known as the pukis or paukis would transform into when on the ground. Only when airborne would it assume its

Medea riding her dragon-drawn chariot.

true reptilian form, during which phase it would emit fire from its tail (reminiscent of the ejderha, a Turkish serpent dragon).

Greek mythology is extensively populated by winged classical dragons. One such example was the dragon of Colchis, who guarded the Golden Fleece obtained from the winged ram Chrysomallus until slain by the hero Jason. Medea, Jason's betrayed wife, made her escape in a chariot drawn by a pair of winged dragons. A similar vehicle also carried Demeter, goddess of agriculture, through the sky. Demeter eventually lent her chariot to Triptolemus, enabling him to travel throughout the world and distribute corn to its inhabitants.

Plutus, god of wealth, was the son of Demeter, but he had been blinded by Zeus so that he would distribute his gifts in an unbiased manner. Happily, however, his sight was restored by a pair of classical dragons in the temple of Aesculapius licking his eyes.

ST. GEORGE AND THE DRAGON OF SILENE

Quite possibly the most famous dragon of all, this swamp-dwelling monster from Silene, Libya, was the voracious maiden-devourer that was supposedly slain by St. George—a 3rd-Century Roman soldier and Christian martyr. Although it is most commonly described and depicted as a

St. George and the dragon, a medieval-style depiction.

mighty winged or wingless classical dragon, this particular specimen has attracted such interest through the ages that many other representations of it also exist.

There are many variations upon the theme of the legend itself too, but according to the most familiar version, the king of Silene and his subjects were being besieged by a monstrous dragon that had crawled out of a huge swamp and was now laying waste to the country—blighting the farmers' crops with its toxic breath, and devouring their livestock with an insatiable appetite until none remained. In order to pacify this foul beast, the king had no option but to demand that each day a child be sacrificed to it. Finally, it was the turn of the king's daughter, Princess Alcyone, to be sacrificed.

Shortly after she had been tied to a stake to await her grim fate, however, a Roman soldier-turned-Christian knight called George, originally from Cappadocia, Turkey, appeared on the scene, and duly entered into battle with the dragon. After a fraught confrontation, George finally overcame his reptilian foe, and subjugated it with the girdle from Alcyone's robe. After leading it back to her father's castle, George beheaded the dragon, freeing Silene from its oppression, before riding away to battle evil elsewhere, eventually becoming a martyr and saint.

In reality, this stirring epic is nothing more than a romantic fantasy, adapted in medieval times from earlier Greek myths, such as Perseus battling Cetus. Even so, this did not prevent it from becoming the most familiar, tenacious dragon legend of all time after it was brought back to Europe by the Crusaders.

A very different encounter between a saint and a dragon featured St. Simeon Stylites (c.390-459 AD), famous for spending 37 years sitting on top of a tall pillar near Aleppo in Syria. One day, his pillar was approached by a huge dragon, whose venomous breath and vile behaviour had decimated every living thing in the area. The dragon had recently been blinded in its right eye when a large branch from a tree fell into it, and now the wounded beast, with head bowed low, wound itself humbly around the pillar's base in silent supplication. The saint gazed down in pity upon the dragon, and as he did so the branch fell out of its eye. The dragon was healed, but no more did it pillage the countryside, contenting itself instead by resting awhile at the pillar's base before making its way to the gate of the nearby monastery. There it worshipped for two hours, and then returned to its den, forsaking ever after its erstwhile malevolence.

THE MUSHUSSU OF BABYLON

Also known as the sirrush, this is probably one of the oldest examples of a wingless classical dragon on record, dating back to the reign of King Nebuchadnezzar II (605-562 BC), among the most esteemed rulers of ancient Babylon. Along with the lion and the bull, the mushussu was one of the three regal beasts depicted in profile upon the magnificent Ishtar Gate of Babylon, erected during Nebuchadnezzar's reign and dedicated to Ishtar, the Babylonian fertility goddess.

Richly adorned with glazed bricks of dazzling cobalt blue and several horizontal rows of these three animal types depicted in realistic form with golden bricks, this arched gateway was a veritable wonder, spanning the processional way between the temples of Ishtar and the sun god Marduk. After the fall of Babylon in c.39 BC, however, it remained buried in the sands for many centuries, until finally rediscovered by German archaeologist Prof. Robert Koldewey during the late 19th Century. Excavations began in 1899, and took several years to complete. It can now be seen in all of its fully-restored resplendence within the Vorderasiatisches Museum in Berlin, Germany.

The mushussu had a long slender scaly body and tail, which was carried almost vertically. By way of symmetry, it also had a long and near-vertical neck, with a typical dragon's head bearing a pair of curled horn-like structures and what may be a single vertical horn upon its brow like a reptilian unicorn. A long forked tongue emerged from its mouth. Its four legs were all sturdy, but the forelimbs were quite different from the hind ones. Unscaled, they terminated in mammal-like clawed paws; whereas the hind limbs' upper regions were scaled, and their lower regions resembled those of a mighty eagle, armed with huge talons.

Of interest is whether this dragon was of the same type as the example referred to in 'Bel and the Dragon'—one of the books

The mushussu or sirrush as portrayed upon the Ishtar Gate. (Vorderasiatisches Museum)

of the biblical Apocrypha. This book tells of how the people in Babylon once worshipped as a living deity a dragon that lived in the temple of one of their gods, Bel—until it was choked to death by Daniel, in order to demonstrate that it was merely mortal, just like themselves, and was therefore not a genuine god deserving of veneration by them.

Another Middle Eastern classical dragon was the akhekhu, possessing a very long, sinuous, serpentine body but also

16th-Century engraving of the dragon in Bel's temple.

equipped with four sturdy claw-footed legs. And the asdeev was a forest-dwelling dragon with white scales and bat-like wings that was eventually dispatched by Rostam, a famous warrior-hero of Persian legend, as one of the seven great labours imposed upon him by the giant demon-like divs after they had

captured him. Also occurring in the mythology of ancient Persia was the ganj, a huge winged classical dragon that guarded vast quantities of treasure, including gold, silver, and precious gemstones, as well as mystical artefacts. It even had a magical jewel embedded within its brow that grew there of its own accord (such stones have been reported from several other forms of dragon too).

THE MAKARA AND ODONTOTYRANNOS —AMPHIBIOUS ASIAN DRAGONS

In the Far East, predominantly within the mythology of India, a notable dragon very different from the familiar Oriental sky dragons occurred. This was the makara.

It was commonly represented as a scaly wingless classical dragon of crocodilian appearance but with a fishtail rather than an unfinned tail, emphasising its amphibious nature. Having said that, some makara descriptions presented a much more exotic, composite beast. These ascribed to it such characteristics as the foreparts and/or the trunk of an elephant, the feet of a lion, the eyes of a monkey, the tusks and ears of a wild boar, and even sometimes the tail of a peacock rather than a fish.

The makara was frequently depicted as the steed or vehicle of various deities—in particular, the river goddess Ganga in Hindu iconography, and, in Vedic portrayals,

Terracotta makara figurine. (Dr. Karl Shuker)

THE PIASA AND OTHER CLASSICAL DRAGONS OF THE NEW WORLD

Perhaps the most unusual classical dragon was the piasa or man-faced dragon-bird of Illinois, USA. In August 1673, Jesuit priest Father Jacques Marquette was travelling along the Mississippi while journeying through Illinois when, looking up at the cliffs towering above both sides of this mighty river at Alton, he was both horrified and fascinated by some huge, extraordinary petroglyphs carved into the face of one of the cliffs.

They depicted a truly astonishing monster, which the local Indians informed him was known as the piasa. In overall appearance, it closely compared with the famous winged classical dragon of European mythology. Boldly adorned in black and red scales all over its body, the piasa had four limbs whose feet were equipped with huge talons. It bore a pair of long antler-like horns upon its head, it sported an extremely

Varuna, the Vedic sea god. A Tibetan variant was the chu-srin, but this was portrayed as an almost exclusively mammalian composite beast, incorporating features from the lion, horse, and stag. Further forms are on record from Thailand and Indonesia.

Another atypical, amphibious form of Asian dragon was the odontotyrannos, indigenous to the banks of the Ganges River. So prodigious in size that it swallowed elephants in a single gulp, this wingless classical dragon had armour-plated scales and bore three separate horns upon its brow. During Alexander the Great's Asian conquests, he and his army supposedly encountered an odontotyrannos, which savagely attacked them, resulting in the deaths of more than 20 men before it too was finally killed. Its carcass was so vast that it allegedly required 1300 men to transport it away.

A naga emerging from the mouth of a makara at the Phra Maha Chedi Chai Mongkol temple, Thailand.

long tail with a forked tip, and two enormous bat-like wings with vein-like markings were raised above its body. But what set the piasa entirely apart from other classical dragons was its bearded face—for in spite of its snarling grimace of fang-bearing teeth, broad nose, and flaming eyes, it was nonetheless the face of a man!

According to the Indians, the piasa had lived in a huge cave in the cliff face and was once friendly to humans—until it acquired the taste for their flesh. Afterwards, it became a bloodthirsty, insatiable killer, but was finally lured within range of the tribe's best marksmen, who severely wounded it with a barrage of arrows, then finished it off with their tomahawks.

Tragically, in c.1856 these wonderful ancient petroglyphs were destroyed accidentally during some quarry work nearby, which caused the petroglyphs to crack and shatter, falling off the cliff face into the river.

Another antlered classical dragon indigenous to North America was the wakandagi. Adding to its deer-related characteristics were its hoofed feet, but this was a water dragon, not a terrestrial one, inhabiting the Missouri River, and it features in the ancient lore of the Mohawk and Omaha nations. A most unusual creature in every way, it could only be seen when viewed through a mist, and would hurl spheres of water at any intruder,

which would burst and flood his boat unless deftly caught and thrown back at the wakandagi.

A more conventional form of classical North American dragon was the gowrow. Wingless, web-footed, green, and unable to breathe fire but armed with a pair of very formidable tusks, a long tail bearing a blade at its tip, and horny serrations running down its back, this extremely large lizard-like beast (up to 20 ft long) occurs in traditional Ozark folklore. A savage predator that wouldn't hesitate to attack and devour anything, or anyone, that it encountered, the gowrow usually inhabited deep caverns or hollows that for obvious reasons were rarely penetrated by any but the bravest or

Piasa—a modern-day depiction at Alton, Illinois. (Burfalcy/Wikipedia)

most foolhardy of travellers. Gowrows hatched from eggs the size of beer barrels, and were initially carried by their mother in a pouch—the world's only marsupial dragon?

One other lizard-like classical dragon of the New World was the kikituk. This huge wingless version features in Inuit folklore.

The gaasyendietha of northwestern North America was a fire-breathing winged dragon of immense stature that lived in rivers and lakes (particularly Lake Ontario), but would also take flight through the sky, spewing forth great gusts of flame as it flew by. According to the mythology of the Seneca nation, when a meteor streaks across the heavens, it is really the gaasyendietha.

Another fire-breathing classical dragon of the New World, but this time wingless, was the epunamun of Chile. According to Mapuche tradition, as documented in Chilean Jesuit priest Father Alonso de Ovalle's *Histórica Relación del Reyno de Chile* (1646), it was a terrible sight to behold, spewing forth flames in all directions, and brandishing its powerful curled-up tail.

A notable South American aquatic dragon was the maripill, which inhabited lakes in northern Patagonia. It was described in Mapuche folklore as resembling a tall, misshapen horse in outline but covered in scales, with a long neck, a very large lizard-like head whose lengthy jaws were crammed with sharp teeth, and long clawed limbs. Its back bore a very sharp, saw-like jagged edge running down its entire length, which it would use to rip cattle apart by running directly underneath their bellies.

THE DRAGONS OF OCEANIA
Dragons belonging to the wingless classical category may well be most closely associated with Europe, but some have been reported far away from that continent. Among the most fascinating yet least-known of these remotely-located classical dragons were those of Oceania.

Australia was home to several very different forms. One of them was a freshwater version known as the kurreah, which inhabited Boobera Lagoon's deep lakes and underground springs in New South Wales. Thanks to its crocodilian jaws, it was sometimes assumed by Westerners to be nothing more than a real crocodile, but was much more than that. Not only was its body very elongate and snake-like, its extremely lengthy, slender tail was prehensile, able to grip anything that it wrapped its tip around. Its feet were broadly webbed, it sometimes sported exotic frills around its neck, was of colossal size, covered in thick scales, and occurred in several different colours, including green and orange. If a kurreah spied anyone swimming in its abode, it would not hesitate to seize the hapless human in its huge jaws and haul him underwater, to drown and then devour him.

Even more ferocious than the kurreah, however, was the burrunjor, a huge wingless dragon named after a remote expanse of Arnhem Land in northern Australia called Burrunjor where it allegedly roamed—and still does today, at least according to local aboriginal testimony. Whereas most classical dragons were quadrupeds, the burrunjor was bipedal, striding purposefully along upon its hind limbs like a carnivorous theropod dinosaur from prehistory, and aboriginal cave art portraying a creature fitting this description still exists in the Burrunjor region. Similar monsters also exist in the folklore of several tribes inhabiting Papua New Guinea.

Another bipedal dragon Down Under was the gauarge or gowargay. Resembling a featherless emu, this pitiless monster frequented water holes. If anyone were foolish enough to bathe in a gauarge's water hole, it would whip it up into a mighty whirlpool and drag the unfortunate bather down into its swirling depths to drown him.

The most famous freshwater dragon in Australia, however, was definitely the bunyip. Although modern reports of mysterious creatures reputed to be bunyips are often likened to dog-headed mammalian creatures, the traditional bunyip of aboriginal lore was a huge aquatic wingless dragon, whose fearsome presence was readily made apparent by its spine-chilling, ear-splitting bellow. Bunyips were fiercely protective of their young, and one famous myth tells of how a whole tribe was transformed into black swans when one of their hunters abducted a young bunyip from its lake and was relentlessly pursued by its enraged mother, in whose wake the entire lake was drawn up, completely submerging the tribe's village.

The mindi was a very specialised form of bunyip, very serpentine in body form, so was sometimes referred to as a bunyip-snake and deemed to be a rainbow dragon too. Of immense size, and possessing magical

The bunyip. (Richard Svensson)

powers according to the ancient lore of the Yarra Yarra aboriginal people, it could readily poison any would-be attacker and also spread disease.

Yet another water dragon of Australia was the oorundoo, native to the Murray River. According to aboriginal legends, this enormous aquatic beast created Lakes Victoria and Albert.

New Zealand's native dragons were (or are?) the taniwha. Looking somewhat like gigantic gecko lizards or colossal tuataras (the tuatara being a unique, primitive reptile surviving only in New Zealand), but bearing a row of long sharp spines along the centre of their back, taniwha are still seriously believed in even today by the Maori people, and are said to have formidable supernatural powers. In 2002, a major highway in New Zealand had to be rerouted because of Maori claims that it would otherwise intrude upon the abode of a taniwha. And as recently as 2012, a similar objection arose in relation to the planned $2.6 billion construction of a tunnel in Auckland, with protestors claiming that this would disturb a taniwha that lived under the city.

Auckland notwithstanding, these formidable dragons normally inhabited dark, secluded localities on land, as well as in large freshwater pools, and sometimes in the sea too, and were reputedly able to tunnel directly through the earth, often causing floods or landslides as a result. Each taniwha was allied to a specific Maori tribe that it protected as long as it received a fitting level of respect and veneration, but it would often attack and devour members of other tribes.

Also present in Maori traditions are the ngarara—giant lizard-like land dragons seemingly resembling monitor lizards (even though these are not known to be native to New Zealand). Various ngarara could assume the form of a beautiful young woman (as could some taniwha).

Very prevalent in traditional Hawaiian mythology is an enormous shiny-black wingless dragon of infamously mercurial temperament known as the moho or mo'o. Like certain dragons of New Zealand but otherwise unlike most dragons elsewhere in the world outside the Orient, the moho was a skilled shape-shifter, normally measuring 10-30 ft long but able to transform instantly if need be into a tiny, inconspicuous gecko-like lizard, as well as a beautiful seductive woman.

Invariably associated with water, the moho was predominantly a guardian spirit deity, protecting individuals or entire families, as well as districts, and specific localities such as fishponds—which if deep enough were frequently inhabited by these dragon deities. Although they would often remain hidden beneath the water, consuming in ecstasy the intoxicating kava root, their presence in such ponds was betrayed if there was foam upon the surface, or if fishes caught there tasted bitter.

Similar dragons were reported from other Pacific Ocean islands or island groups too, including Tahiti and Tonga. Indeed, they were actively worshipped on Tahiti by the royal Oropa'a family. And on Tonga, lizard-like or crocodile-like dragons of prodigious size and lake-dwelling propensity,

Taniwha carving.

reputedly sent by the gods, would seize unwary bathers or women washing items in their lakes, and promptly plunge down into the water with them, drowning their unfortunate victims beneath the surface.

ADDITIONAL APPENDAGES

Several classical dragons reported from around the world possessed multiples of certain body parts—such as heads, tails, and even tongues—that in the majority of dragons were normally represented by just a single example.

A number of different multi-headed classical dragons exist in the legends and lore of Eastern Europe, such as the green-scaled triple-headed slibinas of Lithuania, the fin-bearing balaur of Romania (whose saliva could reputedly engender precious stones), and the sarkany of Hungary. This last-mentioned dragon was unusual among Western dragons in that it was a shape-shifter, existing for much of its time in the guise of a human, albeit an extremely tall, burly, brutish one.

The most spectacular multi-headed dragon from this region, however, was a winged female example from Russia—the goryschche. She boasted no less than 12 heads—until, that is, after having kidnapped hundreds of youths and imprisoned them deep within her subterranean cavern, she was confronted by a local hero, Dobrynya Nikitich. He lopped off 11 of her 12 heads, then successfully slew her following

Knight battling a three-headed
dragon, portrayed in a 19th-Century
fairytale illustration.

a second confrontation after she had abducted the tsar's daughter.

The Albanian bolla had a serpentine quadrupedal body with two small wings, and for its first eleven years of life was almost perpetually asleep—only waking on St. George's Day each year, whereupon it devoured the first human it saw, then fell asleep again. In its twelfth year, however, it developed nine tongues, as well as horns, spines, larger wings, and fire-breathing ability. It was then called a kulshedra or kuçedra, and caused severe droughts unless fed a diet of human sacrifices.

Orochi was an infamous eight-headed, eight-tailed dragon of prodigious size from Koshi, Japan. It had devoured seven of the eight daughters of an elderly couple encountered by the hero Susa-no-wo, brother of the Shinto sun goddess Ama-terasu. He fell in love with their last remaining

Susa-no-wo slaying Orochi, by Toyohara Chikanobu, c.1870s.

daughter, and, vowing to save her from Orochi's lethal jaws, cunningly lured the dragon with eight vats of potent sake, until it was so inebriated that he was able to chop it up into countless pieces.

15th-Century medieval tapestry portraying the dragon of the Apocalypse, left; and also the leopard-headed Beast of the Apocalypse, right.

The biblical dragon of the Apocalypse had seven heads, each bearing a crown, and also possessed ten horns. Scarlet in colour, and winged, it was really the devil in disguise, who battled with the other rebel angels against St. Michael and Heaven's mighty host, until, as chronicled in the Book of the Revelation of St. John the Divine, St. Michael's valiant army overcame them all and cast them down to earth.

Not content with two heads, the very elongate, river-dwelling pal-rai-yuk of Inuit mythology in Alaska also had two tails, three stomachs, and six legs, as well as a row of fearsome-looking spines running along the centre of its back. To ward off attacks from this formidable monster, Inuit fishermen would paint images of it on their canoes.

UNICORN DRAGONS

Whereas most horned dragons, regardless of type, possessed paired lateral horns on their heads, there are a few classical dragons on record that bore a single central horn instead.

These veritable unicorn dragons include: the Armenian svara—a yellow dragon that was not a fire-breather but was extremely venomous, with big fangs and even bigger ears, finally slain by the hero Keresapa; the tcipitckaam—a wingless alligator-like water dragon with a black body, yellow spiralled horn, and very aggressive temperament that inhabited Lakes Ainslie and Utopia in eastern Canada; and the dragua—a big-eared bat-winged quadrupedal dragon from Albanian folklore that inhabited dense forests, lived for up to 1000 years, could become invisible if need be, and couldn't be killed by humans.

DRAGON STONES

One of the most remarkable characteristics reported for classical dragons was the presence of an extremely valuable gemstone, usually set deep inside their skull (but occasionally elsewhere—see Cysat's account below). Possessing a fiery hue and magical, medicinal properties, this jewel was referred to variously as a dragon stone, draconite, or carbuncle. St. Isidore of Seville (c.560-636 AD), a celebrated scholar as well as Seville's Archbishop, included the following account of dragon stones in his hugely-influential *Etymologiae* (the first Christian encyclopaedia):

> It is taken from the dragon's brain but does not harden into a gem unless the head is cut from the living beast; wizards, for this reason, cut the heads from sleeping dragons. Men bold enough to venture into dragon lairs scatter grain that has been doctored to make these beasts drowsy, and when they have fallen asleep their heads are struck off and the gems plucked out.

This interesting lore is very reminiscent of the medieval belief that toads possessed a brilliant gem of unequalled beauty inside their heads. It also recalls afore-mentioned native claims concerning the presence of a diamond in the head of the ninki-nanka, a West African serpent dragon; and the precious jewel supposedly embedded within the brow of the ganj, a Persian dragon.

Rennward Cysat was a 16th-Century town clerk in Lucerne, Switzerland, who penned a detailed chronicle of local events, and which contained a very remarkable story. Cysat claimed that a century earlier, in 1421, a farmer named Stempfli had observed two classical winged dragons flying from Mount Rigi towards Mount Pilatus (home of the venomous dragonet slain by Winckelriedt according to Swiss folklore), and that during this flight one of the dragons had dropped a very strange stone from its abdomen. The dragons' virulent breath had temporarily asphyxiated the farmer, causing him to faint, but after regaining consciousness, he had found this strange stone, which was black, white, and red in colour, and weighed just over 0.5 lb, lying beside him in a pool of blood.

Many mystifying relics like this tend eventually to be lost or discarded, but fortunately the dragon stone became an heirloom within the Stempfli family, passed down and treasured from generation to generation. They claimed that it possessed healing qualities, and to ensure its safekeeping it was eventually donated to Lucerne's Natural History Museum, where it can still be seen on display today. In 1986, two separate scientific institutes in Switzerland conducted an investigation of this cult item, though no samples of it were removed for mineralogical analysis due to its historical significance.

From their studies, the teams were able to discount the once-popular theory that it was a meteorite, and discovered that it was slightly radioactive, but it possessed no magnetic qualities. They concluded that it was most probably a silicate, but its origin remains a mystery—unless it really did fall from a dragon's abdomen! Certain other preserved dragon stones have proven to be fossil ammonites or rare crystal formations.

AERIAL DRAGONS

Aerial or sky dragons came in all shapes and sizes, and have been recorded in many parts of the world. They were united morphologically by their generally serpentine body form and dragon head, however, and behaviourally by spending much of their existence airborne, rarely if ever coming down to the ground.

AMPHIPTERES AND GWIBERS

One of the most common Western forms of sky dragon was the amphiptere, which resembled a serpent dragon in overall form, i.e. possessing a long and limbless ophidian body with the head of a dragon, but it also sported a large pair of bat-like wings. These enabled it to fly above the ground, its body frequently held in a series of vertical undulations. Some amphipteres also possessed a pair of rooster-like facial wattles. Although they ought not to be, amphipteres are often confused with winged snakes—mythical reptiles entirely ophidian in form, lacking a dragon's head—so these latter beasts will also be documented later here, in order to emphasise their fundamental differences.

Several amphipteres have been recorded from Britain. What has been claimed by many chroniclers to be not only the last British amphiptere but also the last British dragon of any kind was described by eyewitnesses as a big-eyed, 10-ft-long specimen as thick as a man's leg, with two tongues inside its toothy mouth, but disproportionately small wings. Sighted twice during late May 1669 near Henham

in Essex, according to some reports it was timid enough not to need slaying, but was simply shooed away into the forest by the Henham villagers.

An amphiptere, from Edward Topsell's bestiary *The Historie of Serpents*, 1608.

In Wales, amphipteres were known as gwibers, and were sizeable creatures, but the largest and most dangerous was the gwiber of Penmachno, in North Wales. Uniquely among gwibers, it was able not only to fly through the air and dwell upon land but also to live underwater. After it even defeated the fearless local hero Owen Ap Gruffydd, this most formidable of gwibers was attacked en masse by a group of Owen's friends, firing a volley of arrows into its sleeping form. Waking to find itself besieged and severely wounded, the gwiber shrieked in pain and rage, then dived into the nearby river to escape. It was never seen again.

Like worms, gwibers loved the taste of milk and would often be discovered suckling cows in the fields. Moreover, if a normal snake should happen to lap up any spilt milk, it would turn into a gwiber.

Intriguingly, there are a few reports from Sweden of lindorms possessing a

A rare winged lindorm from
Sweden. (Richard Svensson)

small pair of forewings instead of forelimbs. Their wings were neither big enough nor strong enough to enable these mighty reptiles to fly, but otherwise they resembled traditional Western sky dragons.

A notable example of sky dragons in classical mythology was the pair of these creatures often portrayed as chariot-harnessed beasts serving the ancient Roman deity Saturn—traditionally the god of agriculture and time, equivalent to Chronos (Kronos) in Greek mythology.

QUETZALCOATL AND QUCUMATZ

Quetzalcoatl was the plumed serpent dragon deity of the Aztecs and Toltecs in Mexico; Kukulkan was its Yucatec Mayan counterpart. This spectacular entity resembled an immense serpent dragon, yet was clothed not in scales but in brilliant emerald-green plumes. It was also gifted with the ability to soar through the sky without wings (although artists sometimes portray it as winged).

In terms of representation, the plumed serpent had a dual existence. Often it constituted a disguise, an alternative form, assumed by Quetzalcoatl, god of the wind and sky. But in some stories it appeared as an entirely separate entity, out of whose enormous mouth stepped Quetzalcoatl in human form.

Saturn in his chariot, drawn by a pair of
Western sky dragons. (Noble, 2002)

A less famous Mesoamerican feathered serpent dragon was Qucumatz, featuring in the legends of the K'iche Mayans, who lived in what is today Guatemala. A two-headed water-linked deity of wind, clouds, and rain, he brought agriculture and civilisation

Quetzalcoatl as the plumed serpent dragon, portrayed in the Mexican Codex Telleriano-Remensi.

to these Mayan people, and also carried the sun across the sky and down into the underworld each day. In addition to his feathered serpent guise, Qucumatz could transform himself into a jaguar, an eagle, and a pool of blood, was sometimes represented by a snail or conch shell, and was associated with a flute made of bones.

ORIENTAL DRAGONS

Fundamentally different from Western dragons in morphology, behaviour, role, and reception by humans were the dragons of China, Japan, Korea, and Vietnam.

In general appearance, these dragons were remarkably serpentiform, with enormously long, elongate bodies (ensheathed in exactly 117 scales), to which, almost as an afterthought, two pairs of rather fragile-looking limbs were appended, plus a lengthy tail with a tufted tip. Their feet possessed long eagle-like claws, and the number of claws present per foot was of great significance. In China, only the imperial dragons had five claws per foot; most

The author's carved Oriental dragon statue. (Dr. Karl Shuker)

others here had four. And the principal dragons of Japan (the tatsu) had only three.

The head of an Oriental dragon was somewhat camel-like in superficial form, but sported a lengthy dragon beard, and sometimes a drooping mandarin-style moustache too. Its eyes glittered like precious stones, and it bore a long pair of branched horns upon its head that resembled a deer's antlers. As for its voice: far removed from the deafening roar or bellow of Western dragons, it was generally likened to the jingling of copper pans.

Japanese dragon, by Katsushika Hokusai.

Whereas Western dragons seldom changed very much in appearance, merely in size, as they matured from a hatchling into the adult form, Oriental dragons underwent a series of very dramatic metamorphoses as they matured—a process spanning 3000 years. After hatching from a brightly-coloured egg resembling a precious jewel, the Oriental dragon began life as a water snake, but in 500 years time it had developed the head of a carp. It was now known as a kiao. After a further millennium had passed, the kiao had acquired the scales of a carp too but the overall form of an elongate dragon, complete with four legs, bearded face, and long tail.

At this stage in its metamorphosis, the dragon was called a kiao-lung or lung, which translates as 'deaf', because its ears were functionless at this time. After another 500 years, however, it had grown its horns, and it was via these, not its ears, that it could then hear. This was now the kioh-lung, the most familiar form of Oriental dragon—but it had still not reached the final stage of its metamorphosis. That stage only occurred rarely, and took a further millennium to attain. By then, the dragon had developed a series of branching wings, and was now termed a ying-lung. Not all dragons successfully completed this extensive series of transformations, however, with most types only attaining certain specific stages and never transforming beyond them.

In Korea, the equivalent form of dragon to China's lung was the yong. In Vietnam, it was the rong, although this Oriental dragon was hornless, and bore a series of small, continuous fins running along the full length of its back. There were four notable types of tatsu or Japanese dragon. The largest version, handsomely striped, was the han-riu. Much smaller was the red-scaled ka-riu. The ruling dragon or dragon king was the sui-riu, which was also the controller of rain. And the fuku-riu was the famous Japanese luck dragon.

Undoubtedly the most grotesque of all Oriental dragons was the t'ao t'ieh. One of the oldest types of Chinese dragon, and the symbol of gluttony in Chinese lore, it had a single head but two separate bodies attached to it. Each body had its own tail, plus two hind limbs and one forelimb (thus making six limbs in total). Equally distinctive was the torch dragon, also called the zhulong or zhuyin. A solar deity, it resembled a gigantic limbless Oriental dragon, and was red in colour, but its face was human. It created day by opening its eyes, night by closing them, and the seasonal winds by breathing.

Many Oriental dragons spent much of their time wafting languorously through the skies, which made it all the more surprising that very few of them actually possessed wings. Instead, they owed their aerial abilities to a special gas-filled bladder-like pouch known as the chi'ih muh, situated on top of their head. Male dragons also derived great power from a large luminous pearl concealed from view by special folds of skin present under their chin or throat.

A rare winged Japanese dragon. (Anthony Wallis)

Some Oriental dragons, conversely, eschewed the skies for the lakes, rivers, and seas. Yet like many other mythological beasts of the Far East (but unlike the great majority of Western dragons), most of them were skilled shape-shifters, regardless of their chosen habitat. Some, in fact, spent much of their time in human guise, inhabiting celestial palaces amid the clouds or in equally magnificent residences far below the water surface in deep lakes, pools, rivers, or seas.

A multicoloured four-clawed Chinese dragon with wings, the ying lung, carved from wood. (Dr. Karl Shuker)

Moreover, many of China's most distinguished human families actually claim direct descent from dragons, and some humans could shape-shift into dragons. A famous example deftly combining these two beliefs was the celebrated hero Lei Chen-Tzu, who originally hatched as a baby from an egg created by his real father, the thunder dragon, then subsequently transformed from a human into an enormous green dragon with wings and tusks in order to rescue his foster father Wen Wang, the god of literature.

Some Oriental dragons, conversely, could change into another kind of animal. Perhaps the most (in)famous example was the great white dragon that lived in a huge pond at Yama-shiro, close to Kyoto. Every 50 years, it transformed into a golden-plumaged bird, the o-gon-cho—the mere sight of which, and also the eldritch sound of its howling wolf-like cry, were very bad omens, reputedly signifying the coming of pestilence and plague.

Oriental dragons existed in every conceivable colour and size, and their specific roles could even be determined simply by observing their precise colouration. For instance, a yellow dragon normally brought good fortune, whereas an azure blue dragon heralded the coming of spring and also (at other times of the year) the forthcoming birth of a very important person, and a black dragon signified impending destruction. In China, and to a lesser extent in Japan and Korea too, for countless generations the influence of dragons has infiltrated and pervaded every aspect of human life and culture. Indeed, analysing ancient Chinese texts, linguist Michael Carr discovered over 100 different dragon types, each with its own specific name, form, and role.

Although Oriental dragons were primarily associated with rain and clouds, they also governed every mountain, river, field, and valley, geological seam of metals or gemstones, and even every weather-related phenomenon, from rainstorms, thunder,

Oriental dragon carved on wooden box. (Dr. Karl Shuker)

Winged snakes from Arabia.

benevolent, they could be notoriously capricious.

Curiously, this ambivalence was directly related to the Oriental dragon's body scaling. Out of its total of 117 scales, 81 of them were infused with yang (benevolent essence), and the remaining 36 with yin (malign essence).

WINGED SNAKES WORLDWIDE

Down through the long ages of retelling legends and folklore, mythological winged snakes have been conflated so extensively with amphipteres and certain other sky dragons that they warrant coverage here, in order to distinguish them satisfactorily from bona fide dragons.

In appearance, they resembled normal snakes except for their wings, which could be leathery and bat-like or feathered and bird-like, and could constitute a single pair or, more rarely, two separate pairs. Winged snakes were most commonly reported from the Middle East, especially Egypt and Arabia. The Egyptian snake goddess, Buto, was often portrayed as a huge winged cobra, and other Egyptian deities represented by winged snakes included Mertseger, goddess of silence, and Nekhebet, goddess of childbirth.

and hurricanes to sunshine, rainbows, and fever-spreading winds, so they were duly venerated by humans in the hope of placating them. For although, in stark contrast to their malevolent Western counterparts, Oriental dragons were generally

In early times, small but highly venomous snakes of many different colours but all possessing bat-like wings reputedly existed in Arabia, and congregated in great throngs upon the trees that produced the much-sought-after frankincense resin. So numerous were they, in fact, that during their springtime migration from Arabia towards Egypt, the very air resounded with their incessant hissing and the unceasing beating of innumerable wings. Happily, however, Egypt's sacred ibises soon decimated these toxic ophidian locusts, devouring them in such vast quantities that none remained.

No such snakes exist there today, or indeed anywhere else. Perhaps their macabre mode of reproduction explains

A winged Alpine snake, depicted and documented in 1723 by Johann Jakob Scheuchzer.

their demise. According to legend, at the very height of passion the female winged snake would bite her unfortunate partner's head off, rather like a serpentine praying mantis. And when the young snakes developing inside her afterwards had attained the required size for emerging into outside world, they would gnaw their way out of their mother's body, chewing through her uterus and gut, thereby killing her in the process.

Much more benevolent was the agathos daimon of Greece and Phoenicia, an invisible winged serpent with a heart-shaped tongue (though how this feature is known about if the creature was invisible remains a mystery!). It would hover unseen over humanity, protecting and bringing good luck, and was formerly venerated in its own temples, receiving offerings of wine from its worshippers.

Asturias and Cantabria in Spain were once plagued by cuélebres—a race of gigantic winged serpents that lived in caves, guarding treasure and imprisoning abducted nymph-like entities known as xanas.

Flying snake from the Sien Mountains.

Devouring cattle for food, and laying waste to the surrounding countryside with their toxic breath, these foul creatures sometimes lived for many centuries, but only when they were relatively ancient did their wings become powerful enough to sustain them in flight. During their lengthy lives, their scales became ever thicker until they were impenetrable to weapons—except on Midsummer Night, when they would temporarily lose their invulnerability. So it was then when heroes sought to slay these winged terrors.

The Basque region was home to a serpent that could fly without the need for wings. This was the herren-surge, a seven-headed form that generally maintained a subterranean lifestyle, except when it emerged above-ground to devour stray cattle and other loose livestock.

The sarkanykigyo was a Hungarian winged snake that began life as a huge swamp-dwelling, pig-devouring serpent known as a zomok. Only when it was fully mature did the zomok develop wings, thus becoming a sarkanykigyo.

Winged snakes were also reported from the Alps, but seldom made an appearance in classical European mythology.

The Sien Mountains of China were supposedly home in bygone times to a very distinctive race of winged snake. Each such snake possessed two pairs of wings, enabling it to fly through the air with its elongate body held almost horizontally, its head and neck raised upwards in a graceful s-shaped curve rather like that of a swan. These winged snakes could sing too, but, again like the swan, the quality of their singing left much to be desired, their voices being likened to the beating of stones. The appearance of these uncanny reptiles was not a good omen, as it normally indicated that a great drought would soon occur in the cities nearest to their mountainous abode.

Another Oriental winged serpent was the uwarbami. This was of such stupendous size that it regularly preyed upon humans, soaring down from the skies to kidnap them, until its reign of tyranny was finally ended for good by the hero Yegare-no-Heida.

According to the traditional legends of the Melanesian inhabitants of the Solomon Islands, all life was created by a race of gigantic winged serpents known as the figonas. The supreme figona was Hatuibwari, who had four eyes in a human head, and four breasts that suckled all of creation.

The crowned winged serpent
from de Passe's *America.*

An interesting North American winged snake, complete with feathered wings and a crown upon its head, was portrayed in Crispijn de Passe's 17th-Century tome *America.*

In Chile and Brazil, there are long-standing traditions concerning a very sinister form of flying serpent known as the piwichen. Resembling a modest-sized snake in overall form but possessing a single pair of wings (sometimes said to be feathered like a bird's) and famed for its loud hissing, this greatly-feared reptile brought instant death to anyone who saw it, after which it would proceed to suck the

blood of its prone victim like an ophidian vampire.

The most remarkable winged snakes of all, however, were undoubtedly the jewel-scaled, plume-winged wonders reportedly still existing in Wales as recently as the 1800s. Marie Trevelyan brought these exquisite creatures to widespread attention in her book *Folk-Lore and Folk Stories of Wales* (1909), and her detailed description of them can be summarised as follows.

An elderly inhabitant of Penllyne, Glamorgan, in southern Wales, who died in c.1900, claimed that when he was a boy the woods around Penllyne Castle harboured plume-winged serpents of exceptionally beautiful form, as if covered in multi-coloured jewels, and some also had rainbow-hued crests. They remained coiled at rest, and sparkled all over when gliding swiftly away to their hiding places. They would fly over people's heads if angered, and sometimes their outstretched wings revealed eye-like markings like the ocelli in the expanded trains of peacocks. He stated that this was no yarn, but all perfectly true, and that he and his uncle had killed some because they were "as bad as foxes for poultry" and were "terrors in the farmyards and coverts"—blaming their extermination upon this undesirable attribute.

Much the same description and claims regarding these creatures' predations upon poultry as again recorded by Trevelyan were given by an old woman whose parents took her to visit Penmark Place in Glamorgan when she was a child. The woman also claimed that in the woods around

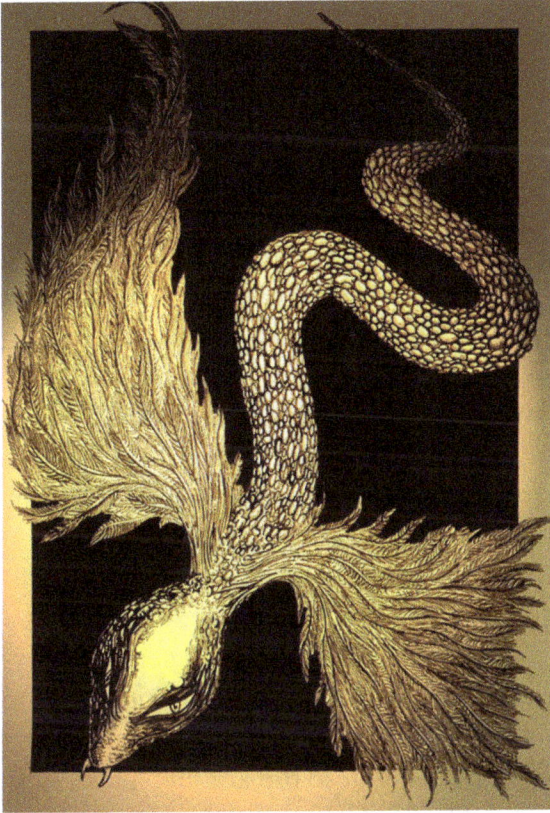

A plume-winged, jewel-encrusted serpent of Penllyne, as depicted in Andy Paciorek's book *Strange Lands*. (Andy Paciorek)

Bewper at that time, there were a "king and queen" of feathered flying snake. Her grandfather, moreover had told her how he and his brother (her uncle) had actually shot one in the woods near Porthkerry Park, close to Penmark, and only killed it after it had put up a fierce fight, beating its wings against her uncle's head. They had then skinned the creature, and the woman claimed to have seen its skin and feathers, but, tragically, this zoologically-priceless relic was discarded after her grandfather died. She also mentioned that when she was a child, the old people claimed that there was sure to be buried money or some object of value near to wherever these winged snakes were seen. In 1812, very similar beasts were also reported in the Vale of Edeirnion, northern Wales.

What could have inspired such astonishing accounts? Remarkably, it has been suggested by certain folklorists and cryptozoologists that these truly exceptional creatures were pheasants. As the ring-necked pheasant *Phasianus colchicus* (the only species in Wales during the early 19th Century) was introduced into Britain back in Roman times, however, it seems unlikely that a species so well-established here by the 1800s could be mistaken for anything as exotic as a plumed serpent with wings.

Even less likely is the prospect that they represented an unknown species of pheasant, one perhaps that was more elongate in form than usual (possessing a longer-than-typical neck and tail perhaps?), thus presenting a vaguely serpentine appearance. For if this were so, there is no doubt whatsoever that many specimens of such a distinctive and extremely eyecatching bird would have been diligently preserved as prized taxiderm exhibits in country manors and museums. Such a striking form of game bird would also have been extensively documented and depicted in countryside magazines or those devoted to hunting and shooting—sports that were extremely prevalent and popular throughout Britain two centuries ago, and obviously it would have been instantly recognised by hunters, poachers, and gamekeepers alike as a mere bird, not some bizarre reptilian entity more akin to the Aztecs' deified Quetzalcoatl!

In short, if such a spectacular, impossible-to-overlook bird had ever inhabited Wales as late in time as the 1800s, it would have been formally discovered and described long before it was ultimately exterminated. Instead, it is conspicuous only by its absence from natural history tomes and from any other wildlife publications, being solely confined to the pages of Trevelyan's book and to scant mentions elsewhere in other works of British folklore.

An avian interpretation of a plumed Welsh winged snake. (Tim Morris)

Equally difficult to explain if these bedazzling beasts were genuinely pheasants (known or unknown species notwithstanding) is their appetite for the farmers' chickens, earning them the reputation of being as troublesome on this score as foxes. Whereas male pheasants can be aggressive, it hardly need be pointed out that they do not feast upon chickens. Ditto for the possibility that they constituted escapee peacocks. And there is no known bird of prey native to Britain (or, indeed, anywhere else in the world for that matter) that matches the multicoloured, glittering appearance described for the thoroughly baffling albeit very beautiful creatures under consideration here.

The same situation also arises when seeking parallels outside the confines of recognised natural history, by venturing forth instead within the more flexible boundaries of zoomythology. An association with buried treasure or similar hoards of riches is of course a familiar theme in dragon myths, but otherwise there is nothing even vaguely comparable between Wales's winged feathered serpents and any other beast of British legend and folklore on record. For not only is their morphology exceptional, so too are their surprising erstwhile abundance and the acceptance of them by the local people as relatively mundane members of the area's fauna.

If only the skin and feathers of the killed specimen noted earlier had been preserved and submitted for scientific examination instead of being discarded. After all, it's not every day that the opportunity to examine the mortal remains of a winged, feathered serpent arises!

NEO-DRAGONS

Neo-dragons constitute a very heterogeneous category of dragon, whose members were not only quite unlike other dragons but also very dissimilar from each another. Indeed, several of the most notable neo-dragons were known from just a single unique specimen. Nevertheless, as they all possessed certain fundamental similarities to dragons, such as the ability to breathe fire, a dragon-like shape, scaling, or behaviour, and clearly shared a common origin with more typical dragons, they definitely warrant inclusion within any classification of these monstrous mythological reptiles.

A salamander carving at the Church of Luigi dei Francesi, near the Pantheon, Rome. (Dr. Karl Shuker)

THE PYRALLIS AND SALAMANDER— FIRE-NOURISHED NEO-DRAGONS

Also known as the pyrausta or pyragones, probably the smallest of all dragons was the pyrallis. In general size and appearance, this smoky-winged neo-dragon resembled a large fly, but had only four legs rather than six, and possessed the head of a dragon instead of an insect's. These tiny but very remarkable creatures could be found flitting like living sparks of flame amid the very heart of the fires in the scorching copper smelting furnaces of Cyprus, but if they ventured out of these fires for even an instant, they immediately died.

A neo-dragon much more famous than the pyrallis for its intimate association with fire was the salamander (not to be confused with the real-life newt-like amphibians that were named after it).

In early days, the salamander was deemed to be a small four-legged wingless dragon with a dog-like head and golden skin brightly patterned with stars, but of such icy, frigid nature that it could douse any fire that it encountered, exuding a strange milky fluid that instantly extinguished all flames. And if one of these deadly creatures entered a body of water, that water would immediately and irrevocably become so toxic that nothing would ever dare drink from it or swim in it again. These claims were being documented by such early classical scholars as Aristotle (384-322 BC) and Aelian (aka Claudius Aelianus, c.175-c.235 AD), so the salamander has a very lengthy pedigree.

It was also said to spit searing, acidic fluid from its mouth, which would not merely put out fire but kill any living thing that it touched. Its bite was lethal too, as was its flesh—nothing could survive after eating the meat of a dead salamander except for the pig. And if anyone killed a pig that had just consumed salamander flesh, and then ate the pig's own meat, they would surely die. According to the Talmud

(Hagiga 27a), however, a person who smeared themselves with the blood of a salamander would thus become immune from harm by fire.

The salamander only came out during heavy rain. As soon as the rain ceased, it would hide away again.

During the Middle Ages, certain types of dragon, including the basilisk and the amphisbaena, underwent a dramatic metamorphosis in folklore and legends. So too did the salamander. From a dragon that repelled and vanquished fire, it transformed into one that was physically sustained by it, and lived unharmed in the very heart of its blazing flames. Italian artist Benvenuto Cellini (1500-1571) claimed to have witnessed such a beast, and even Leonardo da Vinci, normally an accurate scientific observer and commentator, penned the following highly imaginative claim:

This has no digestive organs, and gets no food but from the fire, in which it constantly renews its scaly skin.

The salamander was said to spin cocoons of a furry, fire-proof substance nowadays known as asbestos, but referred to back in those far-off days as salamander wool. Pope Alexander III (reigned 1159-1181) is claimed in the *Buch der Natur* of Konrad von Magdeburg (1309-1374) to have owned an entire tunic woven from this extraordinary material. Fabrics manufactured from salamander wool were apparently washed by being thrown into flames!

By then, however, the salamander had transformed not only behaviourally but also morphologically, because it was now often described as a vermiform beast, rather like a cocoon-spinning silkworm. Moreover, one 13th-Century depiction portrays this protean creature as a winged dog!

A little-known relative of the salamander was the grylio, a mean-spirited creature that delighted in poisoning apples by climbing up into apple trees and licking or rubbing itself against them. If these envenomed apples subsequently dropped off the trees into a well, pool, or some other source of water, it would instantly be rendered undrinkable, irrevocably tainted by the apples' grylio-infused poison.

THE TARASQUE—TORTOISE-SHELLED TERROR OF THE RHÔNE

France has harboured many dragons, but few were stranger than the tarasque and the peluda. Thankfully, each was known only from a single, terrifying specimen.

The tarasque had an impressive parentage, having allegedly been spawned by the immense biblical serpent dragon Leviathan in the Middle East. It subsequently migrated westward into Europe until it found its way to the banks of France's River Rhône, between Avignon and Arles. A most extraordinary, composite neo-dragon—uniting the shell of a gigantic tortoise with the head of a lion, a viperine tail, the scales of a dragon, and no less than six powerful legs, each with a huge bear-like paw and several talons—this monster lay in wait amid woodlands near the town of Nerluc for passing travellers, then burnt them to a crisp with a single blast of its scorching breath before devouring their charred remains.

The tarasque also bestowed the same fiery fate upon anyone brave, or brash, enough to attempt to confront and vanquish it—until that fateful day when St Martha arrived, sailing down the river in a small boat, and docking at Saintes-Maries-de-la-Mer. After Nerluc's inhabitants told her of the tarasque and its grim shadow of oppression, she entered the woodlands in search of it.

So intent on feasting upon its latest victim was the tarasque, however, that it didn't even notice St. Martha approaching until she was standing beside it. Enraged, the dread monster whirled around, about to open its huge jaws and spew forth a scarlet flame of fury, but at that same moment the maiden saint raised two branches before her in the shape of the Cross. Instantly, the tarasque fell back, bemused and strangely quiescent, and from that moment on it was rendered entirely tame, even allowing her to sprinkle it with holy water and bind it with a collar of her own braided hair, before leading the subdued beast back into Nerluc.

St. Martha showed the townsfolk that the tarasque was no longer a threat, and pleaded with them to spare its life, but so many had lost members of their families to its rapacious hunger that they could not control their rage, hurling a barrage of stones and sticks at the cowering dragon until it expired. In recognition of its erstwhile persecution by this monster, Nerluc changed its name to Tarascon, and each year at Whitsun it stages a major festival during which a spectacular life-size model of the tarasque is paraded through the town.

Like so many European myths involving saints confronting monsters, the

Tarasque statue near King René's castle in Tarascon, formerly Nerluc. (Daniel Leclercq)

tarasque legend can be traced back in documented form at least as far as *Golden Legend*. Originally entitled *Legenda Sanctorum* ('Sacred Legend'), this is a collection of somewhat imaginative hagiographies (including one for St. Martha) that was compiled by Jacobus de Voragine, probably in

Edit. Cournand

TARASCON — Procession de la Tarasque

A hand-coloured postcard from 1905, depicting the annual tarasque festival at Tarascon.

c.1260 AD, but was expanded in subsequent centuries by a number of other authors. Moreover, the tarasque was being depicted as long ago as the 1300s, as evidenced by a 14th-Century carved Gothic column capital at the Church of St. Trophime in Arles depicting its tortoise-like carapace in detail, and which can still be seen there today.

THE PELUDA AND THE MIHN—A COUPLE OF HAIRY NEO-DRAGONS

The peluda was also known as the shaggy beast, because whereas most dragons were surfaced in scales, the huge body of this singular creature was profusely covered in long green fur instead. Concealed amid that hirsute mass, however, were countless venom-tipped spines that it could shoot forth like poisoned javelins at anyone bold, or reckless, enough to venture near it. The peluda's long neck and tail were liberally scaled, and its head was that of a huge fire-breathing serpent. Completing its horrifying appearance were two pairs of webbed, turtle-like feet tipped with sharp claws.

Just like the tarasque, the peluda had been spawned in early biblical times, but after being refused entry onto Noah's Ark its amphibious nature had enabled it to survive the Great Flood, and eventually it made its way into the River Huisne, at La Ferté-Bernard in southern France. Here, it lurked on the river banks by day, but at night it not only raided the local farms in search of livestock to devour, but was not amiss to adding women and children to its menu too if the opportunity to do so arose. Happily, this murderous monster was eventually dispatched when a brave hero chopped off its tail—the only portion of its body vulnerable to mortal injury.

The peluda. (Tim Morris)

Less famous yet no less fascinating than the peluda was another hairy neo-dragon—the mihn, featuring in the traditional legends of North America's prairie-dwelling Cheyenne nation. They describe it as an extremely large, lizard-like water monster, but instead of scales it was covered with hair, and bore either one or two horns. Horned, hairy water dragons occur in Sioux traditions too.

THE DRAGON HORSE OF CHINA
One of the most unusual Oriental neo-dragons was the Chinese dragon horse or longma. Traditionally deemed to be the vital spirit of Heaven and Earth, as indicated by its English name it was a curious yet surprisingly effective composite of two very different animals—sporting the body, legs, and hooves of a horse, but the head and scales of a dragon. Some, though not all, dragon horses also possessed a pair of wings, and could walk upon the surface of water without sinking.

Legend has it that eight winged dragon horses pulled the carriage of Emperor Mu of Jin (belonging to the Eastern Jin Dynasty) as he travelled around the world during the post-regency period of his reign (he reigned from 343 AD to 361 AD).

Wooden carving of a dragon horse.
(Dr. Karl Shuker)

It was a yellow dragon horse that emerged long ago from the River Lo to reveal the eight trigrams of the famous divination system known as I Ching. Similarly, a dragon horse rose up out of the Yellow River and gave to the Emperor a circular diagram depicting the yin-yang.

According to the *Imperial Readings of the Taiping Era* (aka *Taiping Yulan*)—a massive 1000-volume, multi-contributor Chinese encyclopaedia dating from the 10th Century—a dragon horse spotted blue and red, covered in scales, sporting a thick mane, and giving voice to a mellifluous flute-like neigh once appeared in the year 741 AD, which was taken to be a good omen for the reigning emperor, Xuanzong of Tang. So fleet-footed that it could cover more than 280 miles without a pause, this wonderful beast had been born to a normal mare that had become pregnant after drinking water from a river in which a dragon had bathed.

During the 7th Century, the Turkestan city of Kucha (now part of China) was visited by the travelling Chinese Buddhist monk and scholar Hsuan-Tsang, who noticed that a lake in front of one of this city's temples contained a number of water dragons. He was informed that they could change their form so as to mate with mares, and that the progeny of this curious crossbreeding were dragon horses, of a fierce, wild nature, and very difficult to tame.

TORTOISE DRAGONS, DRAGON BIRDS, AND OTHER COMPOSITES

The dragon horse is not the only composite neo-dragon on record, as demonstrated by the following varied selection of quasi-dragons, i.e. creatures that were half-dragon and half-something very different indeed!

Modern-day model of a bixi. (Dr. Karl Shuker)

One example was the Chinese bixi or tortoise dragon. Originally, 'bixi' was a term applied exclusively to giant statues of tortoises bearing a pillar or stela upon their backs—a very popular image that inspired many sculptures. In later times, however, such statues began to be given dragon heads rather than typical tortoise heads, and eventually stelae were no longer added. Consequently, the statues being created were now of giant dragon-headed tortoises, and the meaning of the term 'bixi' was expanded so as to apply to this new but very eye-catching composite beast too. It symbolises longevity and heavy loads, and there is a very impressive statue of a tortoise dragon inside Beijing's Forbidden City.

The hai riyo.

Another highly distinctive quasi-dragon from the Orient but with a much longer tradition this time was the Japanese hai riyo or dragon bird. In all but its head, this creature was a large feathered bird, but its head and bearded face, complete with mandarin-style moustache, was unmistakeably that of an Oriental dragon. A highly advanced, eminent creature within the echelons of Japanese dragons, the hai riyo is depicted several times upon the ornamental screens that decorate the esteemed Chi-on-in monastery in Kyoto.

A Chinese version of the hai riyo was the p'eng niao. Again, it was sometimes portrayed as a dragon-headed bird, but could also be depicted as a serpentine dragon with feathery scales and the wings, legs, and feet of a bird.

Stranger by far than the dragon birds of Japan and China, however, was the sachamama or snail dragon of South America. Featuring in early Peruvian Indian legends, it is even claimed to have appeared occasionally in modern times. According to traditional native lore and to various depictions of it on pottery, the sachamama's head was very dragonlike, complete with a fork-tipped tongue emerging from its toothy jaws and a short beard upon its chin. Readily distinguishing it from more orthodox dragons, however, were its two pairs of knob-tipped tentacles. One pair was borne upon its brow, and the other pair sprouted incongruously from its nose.

The sachamama's eyes were very large and black, and a pair of large ears projected backwards from its head. Its long serpentine body was usually said to be uniformly black (though some pottery seen by explorer-journalist Arnost Vasícek depicted it

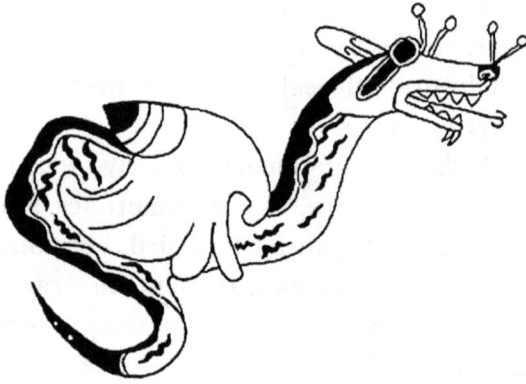

Representation of the sachamama on Peruvian pottery. (Dr. Karl Shuker)

with a rich pattern of curvy lines and squiggles), and it bore a large white conch-like shell upon its back.

The Casinelli Museum in northern Peru's Trujillo region houses several examples of ceramic pots and saucers from Peru's ancient Moche culture (dating back more than 1500 years) that depict this extraordinary creature. It also appears on some 16th-Century examples of Peruvian pottery documented during the 1970s by Vasícek while visiting South America.

Perhaps the most nightmarish neo-dragon was the Australian whowie or yowie (which should not be, but often is, confused with a hairy bigfoot-like man-beast of Antipodean legend also called the yowie). The head of this monstrous composite was that of a huge lizard-like dragon, and its tail resembled a snake's, but its body was an enormous scaly beetle's, complete with six legs!

It would spend the day concealed in a deep cave near the Murray River, but emerge at night and steal into the aboriginal camps nearby to abduct and devour any children or unwary adults that it could find, devouring up to 30 victims each time. One day, however, the tribe set alight a huge pile of brushwood at the mouth of its cave while the whowie was still inside, leaving the fire to burn steadily and fill the cave with smoke, until after a week the whowie was forced to come out, weakened from coughing, its lungs choked with fumes. As soon as it appeared, the tribe's waiting hunters set about it with all manner of weapons until it was killed.

BEARING IN MIND the dragon's unparalleled diversity of form, it will not be surprising to discover that the real flesh-and-blood creatures inspiring much of the lore and legends concerning this fabled beast around the world are just as varied and disparate.

CHAPTER 3:
THE ORIGINS OF DRAGONS

THERE CAN BE no doubt that many stories and legends of dragons around the world were inspired by encounters long ago with a rich diversity of spectacular but entirely real animals. Some of these, such as giant snakes, extra-large lizards, crocodiles, and alligators, are still with us today. Others, such as dinosaurs, pterodactyls, and sizeable prehistoric mammals, exerted their particular influence upon the human imagination not by way of their vital presence but via the remarkable if deceptive appearance of their fossilised remains. Even certain meteorological phenomena may have been viewed as dragons in ages past; and also in erstwhile times the gullible or credulous were often fooled by astutely crafted simulacra into assuming that they had witnessed genuine dragons.

GIANT SNAKES

Some stories of worms and other mighty serpent dragons of mythology unquestionably stemmed from sightings of very large snakes, notably pythons and boa constrictors. Indeed, modern-day pythons actually derive their name from the colossal serpent dragon Python, battled by the Greek sun god Apollo. And boas derive theirs from a monstrous Italian version known as the boas, which sucked countless cows dry of both their milk and their life.

Unlike those of many other snakes, the heads of pythons and boas are very well-delineated, just like those of serpent dragons. Their huge lengths are also very comparable with those reported for some serpent dragons, and in cases where the latter are even bigger, fear-induced exaggeration can certainly be implicated. The world's longest species of snake alive today is the reticulated python *Python reticulatus* of southeast Asia, which regularly attains a total length of up to 20 ft, and sometimes more. One specimen shot in Sulawesi, Indonesia, in 1912 supposedly measured an astonishing 30 ft after having been accurately measured by a team of civil engineers using a surveying tape. Slightly shorter in length but far heavier is the green anaconda *Eunectes murinus* of South America, which can weigh over 440 lb.

Encounters by Western travellers with serpentine monsters such as these, and

Giant python, 1867, from *The Bestiarium of Aloys Zötl 1831-1887*.

their much-embroidered retellings concerning them when safely back home, could have swiftly and very readily engendered all manner of far-fetched yarns concerning vast limbless dragons existing in exotic, far-off realms of swamp and jungle. From this, it would have been only a short step to relocating them in much closer, more familiar lands by imaginative storytellers. And if a few living specimens of such snakes were brought back to Europe from time to time for display in menageries and travelling sideshows, an occasional escape would have been more than sufficient to give rise to further lurid stories and local folklore.

In Australia, moreover, it is even possible that the gargantuan rainbow serpents of Aboriginal Dreamtime mythology were inspired at least in part by a now-extinct genus of gigantic python-like snake known as *Wonambi*. Up to 20 ft long and dying out around 50,000 years ago, it may have survived just long enough to have been seen by the first human settlers in Australia, who would certainly have been awe-struck by the sight of such a huge serpent.

In addition to their great size, another characteristic of serpent dragons linking them to giant snakes is their noxious breath, claimed in various folktales to be so toxic that it can kill livestock and spread infectious diseases. This is clearly nothing more than an exaggerated account of the foetid stench exhibited by the breath of giant meat-eating snakes such as pythons and boas—a quality also reported from other carnivorous animals, such as wolves and big cats.

Equally, the glowing eyes of serpent dragons correspond well with the phosphorescent appearance of the eyes of anacondas and other very large constricting snakes, especially when viewed in dimly-lit surroundings. And it is probably no coincidence that many stories featuring serpent dragons tell of how they were encountered lying upon bright sunlit hills or stretched out in hot wastelands—because this is exactly what one would expect from huge snakes, which sunbathe in order to maintain their body temperature and metabolic rate. Snakes, like other reptiles, are poikilothermic ('cold-blooded'), and are therefore unable to regulate these aspects of their physiology internally, so are dependent instead upon external heat sources to achieve this.

Of course, myths and legends tell of serpent dragons far greater in size than even the largest known specimens of modern-day snake. Whereas this is no doubt due at least in part to exaggeration and storyteller licence as already noted, it is also true (and not merely with snakes) that in earlier times, any exceptional, freakishly large specimens would be favoured targets for hunters looking for spectacular trophies. This selective decimation means that down through the centuries, the maximum size of giant snake species would eventually diminish, surviving in their original stature only within old tales of confrontation passed down in ever more distorted, embellished form from generation to generation, until finally the colossal serpent dragons that never actually existed were born.

Extra-large specimens of known snake species may well explain stories of alleged serpent dragons that were maintained in temples and caves to be venerated as oracles or deities in ancient Greece and Rome. One excellent example was chronicled by Sextus Propertius, a Latin poet-scholar of the 1st Century BC, in his *Elegies*:

Lanuvium [on the Appian Way, roughly 25 miles from Rome] is, of old, protected by an aged dragon; here, where the occasion of an amusement so seldom occurring is not lost, where is the abrupt descent into a dark and hollowed cave; where is let down—maiden, beware of every such journey—the honorary tribute to the fasting snake, when he demands his yearly food, and hisses and twists deep down in the earth. Maidens, let down for such a rite, grow pale, when their hand is unprotectedly trusted in the snake's mouth. He snatches at the delicacies if offered by a maid; the very baskets

An 1825 print of warrior Matsui Tamijiro
battling a giant snake, by Utagawa Kuniyoshi.

tremble in the virgin's hands; if they are chaste, they return and fall on the necks of their parents, and the farmers cry 'We shall have a fruitful year.'

This clearly refers to some very large specimen of snake, and as the maiden was not poisoned even when she placed her hand in its mouth, it is obviously not a venomous species, but almost certainly a python. Similar accounts featuring venerated serpent dragons living in groves in Rome and being fed barley cakes by sacred virgins appear in *De Natura Animalium*, by Rome-based Greek scholar Aelian (c.175-c.235 AD).

Such creatures may well have been African rock pythons *Python sebae*. Common in much of sub-Saharan Africa, specimens of this very sizeable snake species (averaging 15.75 ft long but sometimes exceeding 20 ft) were probably brought back to Rome, because it is certainly depicted in Roman mosaics.

An escapee venerated python may explain the account of a boas form of serpent dragon that, when killed upon Rome's Vatican Hill during the emperor Claudius's reign (41-54 AD), was reputedly found to contain an entire child inside its hugely distended gut.

Generously-embroidered reports of constricting pythons in India are undoubtedly at the root of this fanciful account in Pliny's *Natural History* (c.77-79 AD):

Africa produces elephants, but it is India that produces the largest, as well as the dragon, who is perpetually at war with the elephant, and is itself of so enormous a size, as easily to envelop the elephants with its folds, and encircle them in its coils. The contest is equally fatal to both; the elephant, vanquished, falls to the earth, and by its weight crushes the dragon which is entwined around it.

The basilisk, supposedly inhabiting North Africa's deserts, and whose merest glance could kill, may have been inspired by a very specific and unusual type of real-life snake—the ringhals or spitting cobra. Several species are recognised (some of which are native to North Africa), and all of them incapacitate their prey or potential aggressors by spitting accurately, and from some distance away, a stream of venom into their eyes. Early travellers' tales of this remarkable ability could have been elaborated over generations of retelling into the basilisk's fatal glance.

In addition, the deserts of North Africa and the Middle East are home to a small, harmless species of colubrid snake known as the awl-headed sand snake *Lytorhynchus diadema*. The diadem-like markings upon its head and its yellowish-brown body colouration recall traditional descriptions of the basilisk's appearance and may therefore have helped to inspire belief in the latter.

A spitting cobra brought to England during the early 1600s as an exotic pet or exhibit that later escaped into the countryside could provide a plausible identity for a 10-ft-long serpent dragon reported from St. Leonard's Forest, near Horsham, West Sussex, in August 1614. According to a pamphlet circulated at that time (which is the original source of this report and was subsequently republished in the *Harleian Miscellany*, 1744-1753), it killed two people, two dogs, and several cattle by spitting venom at them, but did not try to devour them.

EXTRA-LARGE LIZARDS

Just as serpent dragons were assuredly inspired by giant snakes, it is highly likely that encounters with real-life mega-lizards, mighty crocodiles, and huge alligators greatly assisted in shaping the myths and legends of wingless classical dragons that appear in traditional lore throughout the world.

Of particular relevance here are the varanids or monitor lizards, widely distributed through Africa, Asia, and Australasia, because some of these are known to attain very impressive sizes. The world's largest living species of lizard is a monitor—the aptly-named Komodo dragon *Varanus komodoensis*, a truly dragonesque creature native to Komodo and a few other small Indonesian islands close by. It can exceed 10 ft in length and weigh up to 150 lb. Even lengthier, growing up to 15 ft long, is Salvadori's monitor *V. salvadorii*, native to New

Guinea and Australia, but this species is much lighter in weight and more slender in build than the Komodo dragon.

Until as recently as 40,000 years ago, however, by which time the first humans settlers in this island continent had already established themselves, Australia was home to a truly gargantuan species of monitor, *Megalania prisca*—estimated to have been up to 23 ft long, to have weighed around

The author alongside a life-sized Komodo dragon model at Chester Zoo, England. (Dr. Karl Shuker)

700 lb, and may have borne a crest upon its head. It would be difficult indeed to find a better inspiration for dragon legends among primitive humans than this monstrous lizard.

Heightening the dragon-like appearance of monitors is their bright yellow-orange tongue, which they rapidly flick in and out of their mouth like a flame. Some large monitor species are also very brightly coloured, comparing well with stories of jewel-scaled classical dragons. Some monitors are aquatic, which could explain reports of amphibious or swamp-dwelling classical dragons. Thus it can be seen how sightings of sizeable monitor lizards could become the basis of some dragon myths.

Indeed, in 1980 a long-feared wingless 'dragon' known to native hunters in Papua New Guinea as the artrellia was finally unmasked as Salvadori's monitor. This discovery followed the capture of a young artrellia that explorers from the Operation Drake expedition visiting PNG at that time identified conclusively as belonging to this very large varanid species.

In addition, certain monitor species will sometimes run on their hind legs over short distances. Consequently, sightings of sizeable specimens of these lizards behaving in this distinctive manner could help to explain the Australian legends of bipedal classical wingless dragons such as the burrunjor and gauarge. The same may well be true regarding a drawing on a cave wall of a comparable monster that was shown by some valley-inhabiting Papuans to traveller David M. Davies, who subsequently documented it in his book *Journey Into The Stone Age* (1969).

In 1173 AD, Castillian traveller Benjamin of Tudela spied many Mesopotamian 'dragons', which he claimed were so infesting the ruins of King Nebuchadnezzar II of Babylon's palace as to render them inaccessible. In reality, these were most probably nothing more than large monitors, which

are native here. One example is the desert monitor *Varanus griseus*, which can attain a total length of almost 6.5 ft; and the widely-distributed Bengal monitor *V. bengalensis*, occurring in southeastern Iraq, can reach 5.75 ft. Both would certainly appear somewhat dragonesque to any non-zoological observer.

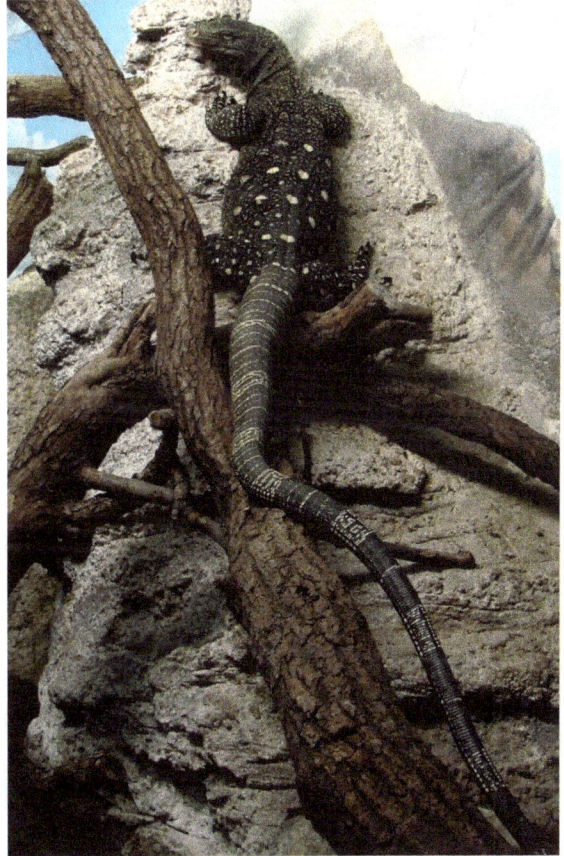

Salvadori's monitor. (Vassil/Wikimedia)

Perhaps it is no coincidence that Asia, home to many different shapes, sizes, and colours of monitors and other notable lizards, is also the birthplace of legends describing a rich variety of quadrupedal dragons. There is even a genus of modestly-sized Asian lizards (containing just over 30 species) with membranous and vividly-coloured wing-like extensions from their ribs that they use to glide from tree to tree. Zoologically, these lizards have been dubbed *Draco* ('dragon'), and are referred to colloquially as flying or gliding dragons. Certainly, they would look just like winged dragons (albeit small ones) to any layman eyewitness.

19th-century engraving depicting two specimens of *Draco* lizard.

CROCODILES AND ALLIGATORS

Probably the nearest living creatures in outward appearance to wingless dragons are the crocodilians—crocodiles, caimans, ghari-als, and alligators. Sometimes exceeding 20 ft long, they are equipped with massive jaws brimming with a fearsome abundance of teeth, an impenetrable plated hide, a lethal array of murderous claws, a long heavy tail that can kill with a single bone-crushing thwack, and a viciously ferocious temperament. Consequently, if in bygone times one of these monstrous reptiles were encountered by a traveller not well-versed in natural history, he would surely have been forgiven for assuming that the very formidable and highly dangerous beast now before him was a real dragon.

Scholars believe that the Nile crocodile *Crocodylus niloticus*, which in antiquity could sometimes be found as far west as southern Europe (having swum across the Mediterranean Sea), inspired many tales of Western dragons, and that the now-endangered Chinese alligator *Alligator sinensis* fulfilled a similar role in the evolution of Oriental dragon mythology. The kurreah may have been inspired by Australia's extinct land crocodile *Quinkana fortirostrum*, which exceeded 15 ft, and had long legs. It died out around 40,000 years ago, but this was after Australia had first been colonised by humans.

There are even some preserved 'dragons' in existence that are undeniably crocodilian in nature. The most notable of these is the 'Brno lindworm' (wingless four-legged classical dragon), of Moravia in the

Czech Republic, which since at least 1608 AD has been hanging suspended from the ceiling of the arched passage leading to the city's town hall, and can still be seen here today. To protect it from the weather, it has been liberally covered in black pitch, but its identity as a crocodile—albeit a very sizeable one, as it measures approximately 15.5 ft long—remains instantly apparent.

The preserved Brno 'lindworm' dragon. (Miroslav Fišmeister)

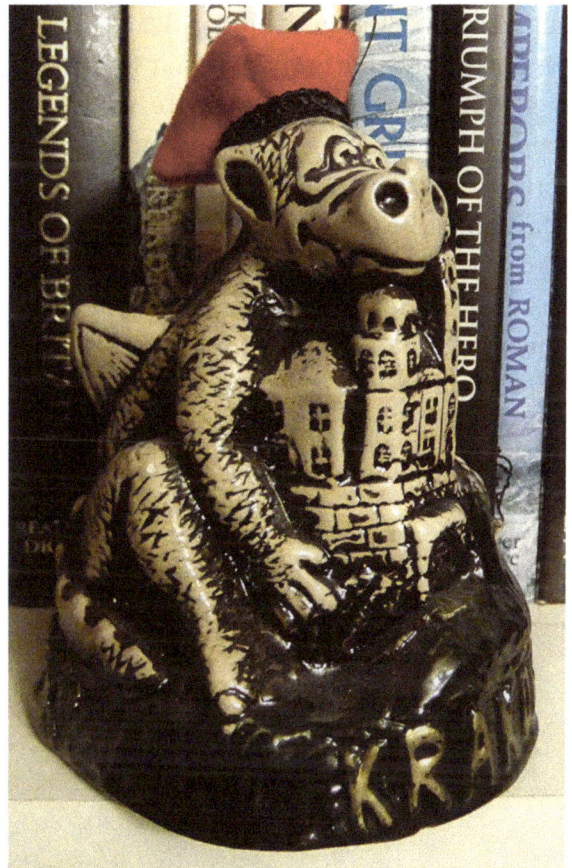

Souvenir ornament portraying Smok. (Dr. Karl Shuker)

According to traditional Moravian folklore, this is the dragon that long ago was said to have slaughtered livestock and even young children in a prolonged assault on Brno (then called Brünn), until it was duped into feeding upon a freshly-killed calf's skin that had been cunningly filled with unslaked lime. After eating it, the dragon was consumed with such a fiery thirst that it drank without pause from a nearby stream until finally the lime reacting with the vast quantity of imbibed water caused the doomed dragon to explode.

The same fate befell Smok, the dragon of Wawel Hill in Krakow, Poland, which had terrorised the city until Skuba, a canny cobbler's apprentice, stuffed a baited lamb with sulphur. (Incidentally, in 2011 a very large species of Polish carnivorous reptile from the late Triassic Period 205-200 million years that may constitute a species of theropod dinosaur was officially christened *Smok wawelski*, in honour of this Polish dragon.)

Another crocodilian dragon once hung from the roof of the cathedral of Abbeville in Picardy, France. And until at least as recently as the early 1850s, what was either a stuffed 6-ft-long crocodile or lizard (monitor?) could be seen suspended in the church of St. Maria delle Grazie, near Mantua, in Italy. Local lore claims that it had been killed in some adjacent swamps in c.1406. More likely is that all of these specimens had been brought back to Europe as unusual souvenirs by returning travellers or crusaders, or even as exotic living specimens for private collections or sideshows.

Hic est Draco ille alatus et quadripes omni aevo memorabilis, quem Deodatus de Gozon Eques Hierosolymitanus, in insula Rhodi eo quo descripsimus stratagemate confecit, qui et ob beneficium in Insulam collatum postmodum Magnus Ord. Magister creatus est.

The Rhodes dragon, imaginatively portrayed in Athanasius Kircher's *Mundus Subterraneus* (1664-1678), long before it was exposed as a crocodile.

According to medieval Greek folklore, during the mid-14th Century a dragon lurked in a swamp on the Greek island of Rhodes, and sustained itself by preying upon farmers' cattle. At that time, the Grand Master of the Knights of Rhodes was a nobleman named Dieudonné de Gozon, and after hearing of the farmers' plight he ignored the previous Grand Master's command not to disturb the dragon, and confronted it with a pack of specially-trained hounds. The dragon was duly slain, and its head was hung from one of the seven gates of Rhodes's principal town. There it remained until as recently as a century ago, when a biologist observed it and soon pointed out that it was actually the skull of a large crocodile.

A similar case dating from much the same time period featured the Holy Roman Emperor Charles IV (1316-1378), a keen collector of saintly relics. One of these was what he believed to be the skull from the dragon slain by St. George. In reality, however, it was from a crocodile.

In 1405, an extremely large, wingless classical dragon with saw-like teeth, a thick armour-plated hide, and a very long, powerful tail killed a shepherd and devoured many of his sheep near the town of Bures and village of Wormingford on the Suffolk-Essex border. Eventually, after being attacked not only by a posse of vengeful arrow-shooting villagers but also by Sir Richard Waldegrave, a local knight, the dragon fled and hid within a great marsh, and was not seen again. It seems very likely that this was a crocodile that had escaped from a menagerie or travelling sideshow. Indeed, local lore claimed that it had broken free from

19th-Century painting of a mosasaur with two ichthyosaurs, by Heinrich Harder.

the royal menagerie in the Tower of London and had journeyed to the River Stour via Essex, and even called it a 'cockadrille'.

Perhaps the most intriguing of proposed connections between the crocodile and the dragon concerns the mighty biblical sea-dragon Leviathan. Several identities have been suggested to explain this stupendous marine monster, including a whale, a shark, and even a surviving prehistoric sea-lizard known as a mosasaur, which was related to the monitor lizards of modern times.

The most popular contender, however, is the Nile crocodile, whose scaly elongated body, glittering eyes, powerful neck, and numerous sharp teeth certainly compare favourably with those characteristics described for Leviathan. In contrast, it conspicuously lacks the latter beast's smoking nostrils (which have been claimed by some scholars to be an allegorical description of a whale's blow-holes), as well as its fins, and it is a freshwater species, whereas Leviathan was a sea dweller. Even today, therefore, the jury is still out on the question of Leviathan's zoological nature—always assuming, of course, that it ever existed in the first place!

SALAMANDERS AND NEWTS

It seems highly likely that the fire-dispelling salamander of legend, which in medieval times underwent so dramatic a reversal of its nature that it became instead a creature sustained by fire, was based upon sightings of its real-life newt-related amphibian namesakes. Certain brightly-marked species, in particular the well-known fire salamander *Salamandra salamandra* of central Europe, closely resemble descriptions of this incombustible neo-dragon. Furthermore, they live in or around damp mossy tree trunks, or under large stones, and if in times gone by a log containing one of these creatures happened to have been thrown onto a fire, the unexpected emergence of so distinctive an animal may well have inspired stories of a fabulous beast that thrived amid the flames.

Real-life salamanders also secrete a toxic skin fluid, just like their mythical counterpart.

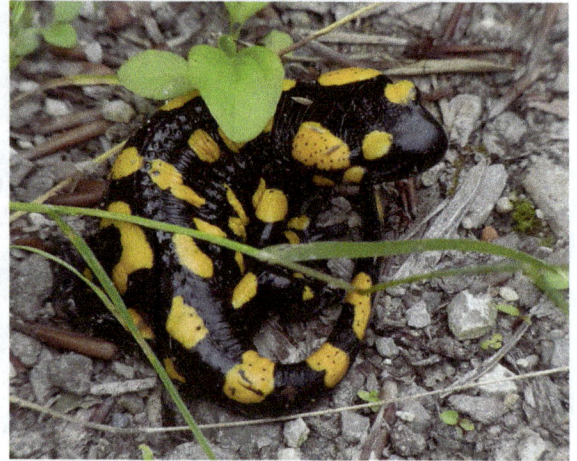

A fire salamander. (Pixabay)

Even the latter beast's original fire-extinguishing capabilities have precedents in the natural world. In a short *Herpetological Review* report from 1997, M. R. Stromberg revealed how, during a prescribed burn on a wooded hillside in California, he had seen two California newts *Taricha torosa* walking through burning

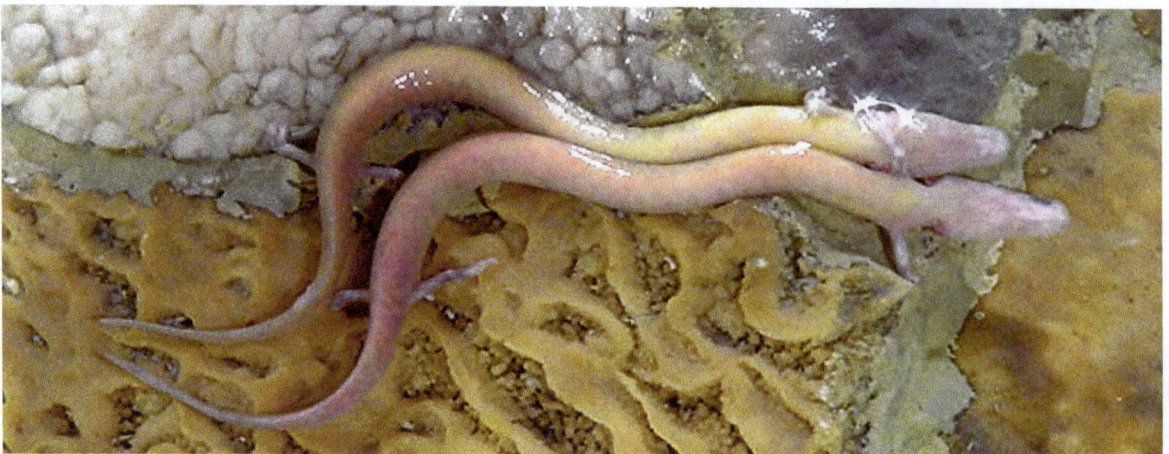

Olms in Postojna Cave, Slovenia. (Boštjan Burger/Wikipedia)

leaves, but remaining protected from the heat and flames by virtue of slimy, foaming skin secretions covering their body and forming a crust. When Stromberg examined these newts, he found no blisters or skin discolouration.

It is not difficult to comprehend how observations of this kind made in much earlier times by non-scientific eyewitnesses could have given rise to imaginative claims concerning the mythical salamander's supposed fire-dispelling nature.

Slovenia in eastern Europe has a dragon-rich mythology, which includes the belief that small white baby dragons are sometimes washed out of their mighty parents' subterranean domain during heavy rains. In fact, such creatures are indeed brought to the surface here on occasion during flooding, but they are not infant dragons. Instead, they are olms—an extraordinary species of serpentine salamander, highly degenerate in form, with regressed eyes, external gills, and tiny limbs.

Known scientifically as *Proteus anguinus*, the olm normally spends its entire life underground in inky black lakes inside caves, where vision is superfluous.

SEX-CHANGE CHICKENS AND OTHER BIRDS

Thanks to its rooster-like coxcomb, wattles, feathers, and crowing cry, the cockatrice was a very unusual dragon, and, as befitting this, it may well have had a comparably odd origin in the real world. Sometimes, an otherwise normal hen develops an internal tumour that stimulates the development of male hormones. These in turn induce the development of male secondary sex characteristics—namely, the rooster's coxcomb, wattles, crowing cry, and even on occasions its plumage. Yet it still lays eggs. In bygone, superstition-laden times, the mere sight of such a bizarre curiosity would have been enough to initiate imaginative fear-laden tales of the dreaded cockatrice.

Cockatrice, by Friedrich Justin Bertuch.

The premier dragon of Central America is assuredly the magnificent green-plumed aerial serpent dragon into which Quetzalcoatl, god of the skies in Aztec and Toltec mythology, transformed when flying through the heavens. As suggested by his name, the inspiration for his wondrous alter ego is an exquisite bird known as the resplendent quetzal *Pharomachrus mocinno*, native to Mexico and Guatemala.

Garbed in brilliant emerald plumes on its upperparts and scarlet feathers on its underparts, the male quetzal is rendered even more flamboyant during the breeding season, when it grows a quartet of extremely lengthy green tail feathers, each up to 3 ft long (i.e. twice the length of the bird's entire body). When it flies, these elongate plumes undulate behind it in a very serpentine manner, so that observers could be readily forgiven for thinking that they had witnessed the passage of a plumed serpent. The quetzal also inspired the K'iche Mayans' feathered serpent, Qucumatz.

Male resplendent quetzal,
1890 chromolithograph.

OCTOPUSES AND SQUIDS

Certain depictions of the Lernean hydra on ancient Greek pottery were quite evidently inspired not by a reptilian dragon, but rather by either an octopus or a squid. Both of these multi-tentacled cephalopod molluscs are common in the seas off Greece and its islands, and it is easy to understand how, after seeing a captured specimen on land flailing its tentacles about its large bulbous body, the legend of a monster with numerous necks could have arisen.

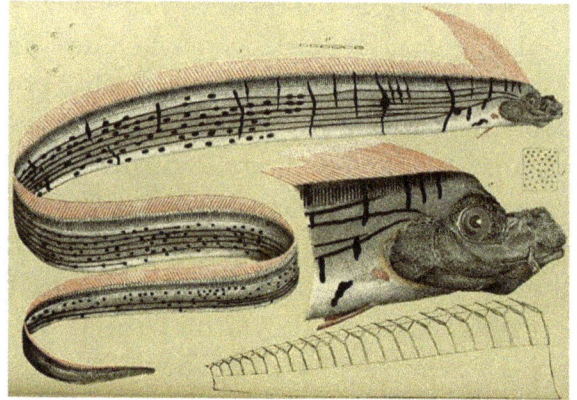

19th-century illustration of an oarfish.

OARFISH

Known scientifically as *Regalecus glesne* and of worldwide distribution, the oarfish is an extraordinary species of highly-elongate marine fish that can measure up to 56 ft long (making it the world's longest species of bony fish), and greatly resembles an enormous snake with a dragon-like head bearing a conspicuous red crest. Occasionally, one of these remarkable fishes is found beached, and when viewed in this state by non-scientists it has sometimes been thought to be a monstrous sea serpent or

The famous photograph of a supposed nagini, clearly an oarfish.

dragon from the deep. Some mythological tales of such beasts are likely to have been based upon early observations of stranded oarfishes. In particular, the crested serpent dragons that came up out of the sea to strangle Laocoön and his sons compare well with this species.

Allegedly seized from the Mekong River by the American Army in Laos on 27 June 1973 during the Vietnam War, a supposed queen naga or nagini is depicted in a famous much-reproduced photograph that is often seen displayed as a curio in tourist bars, restaurants, markets, and guesthouses around Thailand. However, the creature in question is visibly recognisable as a dead oarfish, held up for display by a number of men.

Moreover, it is now known that this oarfish specimen, measuring 25.5 ft long, was actually found not in Asia at all, but off the coast of Coronado Island, near San Diego, California, by some U.S. Navy SEAL trainees in late 1996, and those are the men who are holding it.

PARASITES

Occasionally, a chicken egg when cracked open will be found to contain a small, living worm-like creature. In bygone times, such eggs were referred to as basilisk eggs, because the more superstitious-minded firmly believed that their vermiform contents were young basilisks. In reality, however, they are

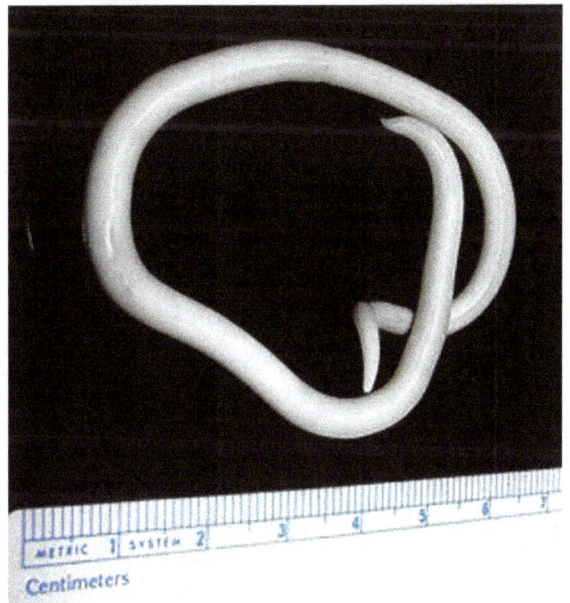

An ascarid roundworm. (CDC)

merely a species of ascarid endoparasitic roundworm (nematode), *Ascaris lineata*, which can grow to a few inches in length.

Normally inhabiting the chicken's gut, one of these worms will sometimes find its way into the bird's reproductive system (which in birds shares a common external opening, the cloaca, with the gut) and thence to its oviduct. Here, the worm becomes incorporated into the albumen of an egg—which, when laid, imprisons the living worm inside itself, until broken open to eat by some unsuspecting diner, whereupon the worm wriggles out!

PREHISTORIC FOSSILS

Some dragons seem so bizarre in shape or form that it is impossible to identify them with any type of animal alive today. There is a very good reason for this—in many such cases, the animals upon which these dragons are based are, indeed, not alive today, but they were once, long ago, in our planet's far-distant prehistoric past.

It is well-documented that some dragons originated from attempts by humans living in the pre-scientific age to make sense of the multifarious array of fossilised animal remains observed by them in a wide variety of locations—in rocks, emerging from cliff faces, washed up or exposed by the sea, dug up out of the ground, or disinterred underground in mines.

Take, for instance, the dragonet of Mount Pilatus. According to medieval legends and documented in German Jesuit scholar Athanasius Kircher's extensive geological tome *Mundus Subterraneus*

(1664-1678), a released convict named Winckelriedt successfully slew on Switzerland's Mount Pilatus a relatively small but exceedingly venomous winged dragon, or dragonet, that had been terrorising the nearby village of Wyler. It is surely no coincidence that the rocks in this same location are rich in well-preserved pterodactyl fossils—and to someone not versed in palaeontology, the fossilised remains of a pterodactyl would look very like what one might expect the skeleton of a small dragon to resemble.

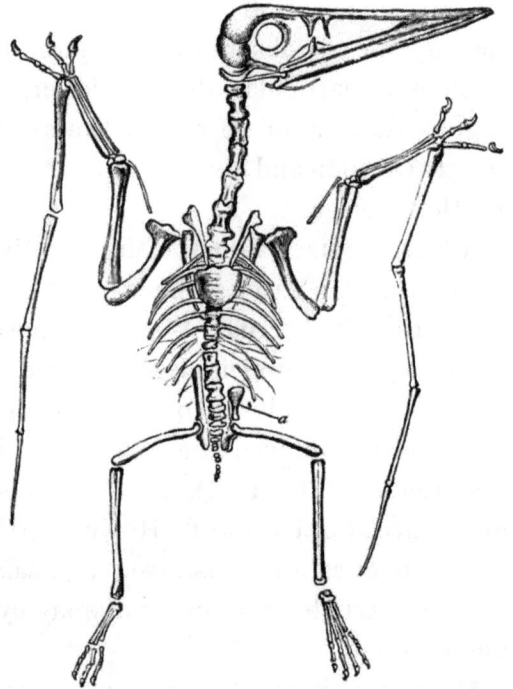

A very dragonesque pterodactyl skeleton.

One of Austria's most famous public sculptures is an exceedingly ornate dragon-shaped fountain in Klagenfurt, created in 1590. The fountain was inspired by the

Klagenfurt dragon fountain. (Johann Jaritz/Wikipedia)

skull of a supposed virgin-devouring dragon that had been slain by two brave men in the vicinity of Klagenfurt in 1335. The skull is still retained in the city's museum, but when examined in modern times, however, it was shown to be that of an Ice Age woolly rhinoceros!

From an illustration prepared of it in 1673 by Johannes Hain, an alleged dragon skull discovered in a cave in the Carpathian Mountains is readily identifiable as that of the extinct cave bear *Ursus spelaeus*. And Drachenhöhle ('Dragon Cave'), a large cave near Mixtnitz, Austria, that was once believed to be inhabited by dragons, has been found to contain the remains of approximately 30,000 cave bears.

Down through the centuries, and even today, one of the staple items sold in the pharmacies and traditional medicine markets of China, and usually found in great quantities inside large glass jars there, have been so-called dragons' teeth (lung-chi). These items are widely believed to have medicinal and oracular properties—beliefs that date back more than 3000 years. In reality, they are the fossilised teeth of a vast assortment of prehistoric mammals (but especially those of a three-toed horse called *Hipparion*) that have been dug up out of fields or the earth by peasant farmers. It was while idly viewing the teeth within one such jar inside a Chinese apothecary shop during 1935 that German-born palaeontologist Prof. G. H. Ralph von Koenigswald made a sensational but totally unexpected discovery.

Among these teeth were some that von Koenigswald readily recognised to be human-like, but they were huge—five or six times bigger than those of ordinary humans. It was as if he had found the teeth of a bona fide giant or ogre! Other teeth of this same type were subsequently obtained, and they all proved to be from a hitherto-unknown species of prehistoric giant ape, the largest ever known, which was aptly christened *Gigantopithecus*.

Fossil shark tooth, often mistaken in earlier days for a petrified dragon's tongue. (Dr. Karl Shuker)

Prehistoric mammal teeth are not the only items of dentition that were once erroneously assumed to have originated from dragons. In bygone times, various large, triangular, serrated fossils, some measuring several inches long, were often deemed to be the petrified tongues of dragons and were referred to as glossopetrae ('stone tongues'). In 1616, however, Italian scholar Fabio Colonna documented their true identity—the fossil teeth of prehistoric sharks.

In Germany, fossil ammonites—those familiar coiled shells resembling rams'

horns but derived from ancient squid-related molluscs—were once thought to have come from dragons and thus were termed drake stones ('drake' being a German-derived term for 'dragon'). According to writer Georg Henning Behrens in 1703, farmers from the Harz Mountains would place these objects into a pail if one of their cows had stopped giving milk, because they genuinely believed that their presence in it would somehow induce the cow to start providing milk again.

A very different kind of dragon stone can sometimes be found in Japan. Sinuous in form, these are crystalline stones of natural occurrence, translucent reddish-yellow in colour, but are believed by some to be fossilised serpent dragons of the worm variety. Due to their rarity and dragon-like appearance, they are greatly prized by Asian (especially Indonesian) shamans for their supposed magical properties, and sell for sizeable sums of money.

Chinese mythology is replete with myths and stories featuring dragons of numerous forms, and China is a land that is equally prolific in terms of the quantity and diversity of dinosaur fossils unearthed through the centuries. Before their finders recognised these fossils' true identity, it was understandable that they should assume they were the earthly remains of dragons.

Indeed, as far back as 1916, in a *Scientific American* article, fossil finder J. O'Malley Irwin had proposed that some traditional Chinese, and also Japanese, dragon lore may well have arisen from early

Arietites ammonite cast, Monmouth Beach, Lyme Regis. (Dr. Karl Shuker)

southern Russia, northern China, and the Gobi desert. And the modern Chinese word for dinosaur, 'konglong', translates as 'terrible dragon'.

Even today in China, the bones of fossil mammals are still popularly collected and sold whole as dragon bones (lung-gu) at about 2 yuan ($0.33) per lb, or in ground-up form as dragon-bone powder (which is then sometimes added into soup), and sold for medicinal purposes in Chinese drugstores. Among the wide assortment of ailments these items are said to cure, according to traditional lore here, are dizziness, bone fractures, and cramp.

In North America, Native American traditions of hairy, horn-bearing water dragons such as the mihn may well have been inspired by discoveries of fossils or even preserved specimens of woolly

discoveries of fossils from huge sauropod dinosaurs here, including some stupendous skeletons of an Oriental sauropod that Irwin himself had encountered while he and his wife had been exploring Shen K'an, a very big cave on the Yangtze River's right bank. Moreover, dinosaur eggs were often claimed to be the preserved eggs of dragons, as too were those of *Struthiolithus chersonensis*, a prehistoric ostrich from

mammoths and mastodons. This fascinating concept was explored in detail by natural history folklore specialist Dr. Adrienne Mayor within her extensively-researched book *Fossil Legends of the First Americans* (2005).

As for the fossilised remains of prehistoric marine reptiles such as ichthyosaurs and plesiosaurs: when these were revealed in cliff faces by sea erosion in pre-scientific

Draco Aethiopicus alatus.

Bestiary image of a fake taxiderm dragon.

times, they were confidently asserted to be the mortal remnants of giant antediluvian sea dragons. Moreover, according to local legend, a dragon known as Blue Ben, frequenting Kilve in Somerset, was regularly ridden through the fires of Hell by the devil, and so he cooled off afterwards by bathing in the sea, until one day he slipped off a rocky causeway into some very deep mud and drowned. However, his skull was later discovered, and is still on display in Taunton Museum today, but observation of it will soon reveal that this alleged dragon skull is in fact that of a fossil ichthyosaur.

FAKE DRAGONS

Stuffed crocodiles and lizards are not the only items that have been put forward as supposed proof that dragons really did exist. A surprising number of wholly fraudulent yet skilfully-manufactured 'pseudo-dragons' have also come to light down through the centuries.

One notable category of pseudo-dragon is the Jenny Haniver. This very imposing exhibit, often four-limbed with a snarling crested head, long tail, and an abundance of fins, wings, and other eyecatching accoutrements, is created using a dead ray, skate, or similar fish that has been dried, and then cunningly modified.

So here, for anyone who may wish to make one for themselves, is a 'recipe' for creating a typical Jenny Haniver, as presented by Gilbert P. Whitley of the Australian Museum in Sydney, within an article from 1928 published by the museum's own magazine:

This is done by taking a small dead skate, curling its side fins over its back, and twisting its tail into any required position. A piece of string is tied round the head behind the jaws to form a neck and the skate is dried in the sun. During the subsequent shrinkage, the jaws project to form a snout and a hitherto concealed arch of cartilage protrudes so as to resemble folded arms. The nostrils, situated a little above the jaws, are transformed into a quaint pair of eyes, the olfactory laminae resembling eyelashes. The result of this simple process, preserved by being coated with varnish and perhaps ornamented with a few dabs of paint, is a Jenny Haniver. . . . The front aspect of the finished article is really the under surface of the skate, whose back and true eyes are hidden by the curled pectoral fins.

Another good example of dragon-mimicking pseudo-morphology using an adroitly-modified dried skate or ray was the preserved amphiptere depicted in 1558 within the fourth volume of Conrad Gesner's monumental work, *Historiae*

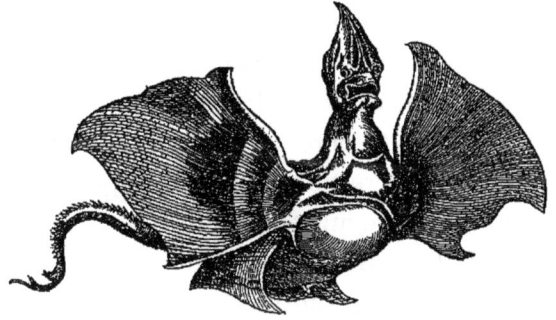

Bestiary image of a Jenny Haniver.

Animalium. Gesner, however, was aware of this specimen's true nature, outlining in his documentation of it the process by which such fakes were created.

Even more marvellous in its own deceiving manner, however, was the hoaxed hydra that was removed from a church in Prague in 1648 and subsequently owned by Johann Anderson, the Burgomaster of Hamburg. So spectacular was this preserved wonder that Anderson even rejected an offer of 30,000

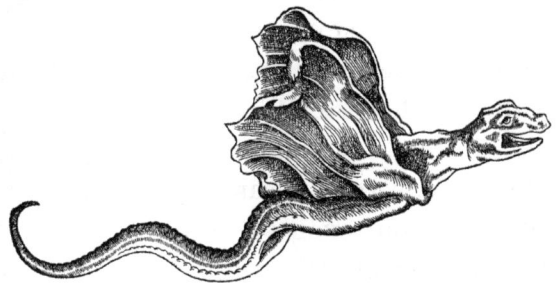

The fake amphiptere depicted in Gesner's *Historiae Animalium*.

thalers for it from Frederick IV, king of Denmark. In basic form, the hydra resembled a standard lindorm, sporting a long tail and sturdy scaled body but only two limbs and no wings. Instead of just a single neck and head, however, it boasted no less than seven of each, with all of the necks emerging from a common base.

Yet despite the hydra's extraordinary appearance, its perceived monetary value eventually decreased, until by 1735 negotiations had begun for its sale at a mere 2000 thalers. Before these could be completed, however, eminent naturalist Carl Linné (who subsequently Latinised his name to Linnaeus) examined this celebrated specimen, and exposed it as a fraud. The heads, jaws, and feet were those of weasels, and a series of snake skins had been pasted all over its body.

Linnaeus speculated, however, that this exhibit had probably been created not by wily vendors to sell as a supposedly genuine hydra to some unwary buyer for an eye-watering sum of money, but rather by monks as a representation of the seven-headed dragon of the Apocalypse with which to chastise and terrify disbelievers. Yet whatever the reason, the result was outstanding, but even so, once this hoaxed hydra's true nature had been revealed by Linnaeus, the deal for its sale fell through, and shortly afterwards the hydra itself vanished—never to be seen again.

In January 2004, media sources worldwide carried reports concerning the remarkable discovery by David Hart of what appeared to be a perfectly-preserved dragon foetus sealed inside a very large jar of formaldehyde, found inside the garage of his home in Sutton Courtenay, Oxfordshire. According to Hart, he had also found with it some documents alleging that this extraordinary specimen had been offered to the Natural History Museum in London back in the 1890s by some German scientists hoping to discredit Britain's premier zoological institution. After the museum had rejected it, however, it had been saved from destruction by Hart's grandfather, who had worked as a porter there; he had retained the specimen ever since.

Depiction of the hoaxed hydra of Hamburg in Albertus Seba's *Cabinet of Natural Curiosities (Vol. 1),* 1734.

In reality, however, as Hart later admitted, it was all a skilful, highly-successful publicity stunt in order to generate attention for his friend Allistair Mitchell's then-unpublished children's novel, *Unearthly*

History (authored by Mitchell using the pseudonym P. R. Moredun). As a result of the stunt, Mitchell won a major publishing contract for his book. The dragon foetus had been prepared by a team of expert model-makers, and the jar had been produced by a specialist glass-blowing studio.

Published in January 2013 by the scientific journal *Palaeontologia Electronica*, a paper written by biologists Phil Senter and Pondanesa D. Wilkins from North Carolina's Fayetteville State University revealed that the skeleton of an alleged dragon from near Rome, and depicted clearly within an engraving by Dutch civil engineer Cornelius Meyer in 1696, was a skilfully-constructed composite creation, i.e. a fake or gaff.

From scrutinising the engraving, the authors readily identified the true, and very disparate, nature of this skeleton's various components. Namely: a domestic dog's skull; the lower jaw of a second, smaller domestic dog; a bear's forelimb (used as the dragon's hindlimb); the ribs of a large fish; a sculpted fake tail; and a pair of fake manufactured wings. Attached portions of skin adroitly hid the junctions between these varied body parts.

A preserved two-legged, two-winged 'dragon' that was a prized specimen within the natural history collections of Cardinal Maffeo Barberini (elected as Pope Urban VIII in 1623) was subsequently revealed to be a forgery also. So too was a mounted specimen of a two-limbed wingless 'dragon' owned by eminent Italian naturalist Ulisse Aldrovandi (1522-1605).

Cornelius Meyer's sketch of an alleged dragon skeleton.

Cardinal Barberini's fake dragon.

Aldrovandi's counterfeit dragon.

METEOROLOGICAL MONSTERS

Some dragons were born not of zoology but rather of meteorology. Scorching through the sky in fiery splendour, these particular dragons were categorised as fire-drakes (fire-breathers), and their appearance, albeit fleeting, brought terror to superstitious observers during the Middle Ages. Even so, their true nature as visual meteorological phenomena was recognised at least as long ago as the reign of Elizabeth I (1558-1603), during which time a chronicler writing in a major tome entitled *Contemplation of Mysteries* stated:

> The flying dragon is when a fume kindled appeereth bended, and is in the middle wrythed like the belly of a dragon; but in the fore part, for the narrownesse, it representeth the figure of the neck, from whence the sparkes are breathed or forced forth with the same breathing.

Furthermore, in modern times researchers have confirmed that many early accounts of fiery sky dragons can be readily identified as sightings of aurorae, which must have seemed truly incredible, even supernatural, to medieval non-scientific observers.

FLYING THE FLAG FOR DRAGONS

Perhaps the most unusual source of inspiration for dragon myths is the plethora of dragon-emblazoned standards, flags, and other banners that were popularly carried aloft on high poles as the symbol of a cohort (500 soldiers) in the Roman army. What made these particularly animate in appearance was the fact that rather than being flat expanses of material, they were created as three-dimensional windsock-like structures, which twisted, writhed, and even hissed ominously in the wind, just like living creatures! In later ages, dragon kites were manufactured that were even equipped with lighted torches emitting fire and smoke as they soared overhead. Consequently, illiterate peasants witnessing their approach and unfamiliar with such exotic devices may well have mistaken these artificial dragons for the real thing.

CHAPTER 4:
THE UNNATURAL HISTORY OF DRAGONS

CRYPTOZOOLOGY is the scientific investigation of creatures whose existence or zoological identity has yet to be formally recognised but which are apparently well known to local people sharing their domain. Judging from eyewitness reports and other anecdotal evidence, some of these elusive mystery beasts (or cryptids, to give them their correct term) bear more than a passing resemblance to various types of dragon from traditional legend and folklore. Indeed, some cryptozoological researchers have speculated that if such creatures genuinely exist, they may even have been the origin or inspiration for certain dragons and myths associated with them—as now revealed via the following selection of putative dragon-engendering cryptids from around the world.

THE TATZELWORM—
A LIVING LINDORM?

Long reported from the Swiss, Austrian, and Bavarian Alps, the tatzelworm is described by eyewitnesses as resembling a fairly sturdy, elongate lizard, about 3 ft long, covered in brown scales, and some-

times with four short stumpy legs but more often only two, near the front of its body.

No such reptile is known to science from this geographical area, but the tatzelworm's description compares closely with the morphology of the mythical lindorm—a dragon type traditionally reported from the mountains of northern and central Europe. So could there be a rare or highly elusive,

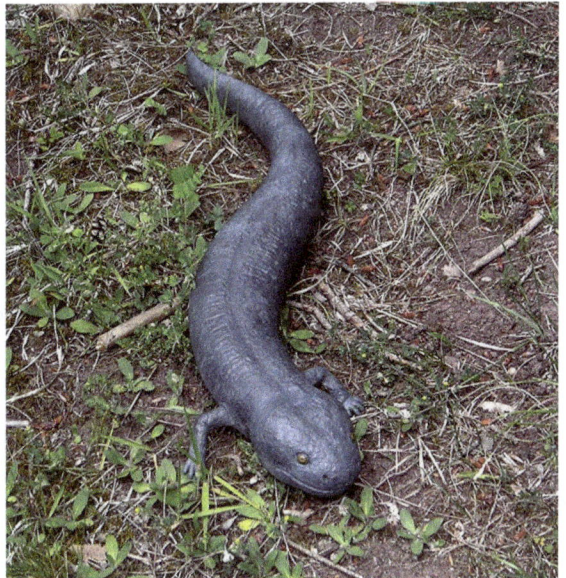

Model of a typical two-limbed
tatzelworm. (Markus Bühler)

Sculpture of a four-legged tatzelworm at Salzburg Museum, Austria. (Ivan Mackerle)

still-undiscovered species of large lizard native to the Alps that gave rise to legends of the lindorms?

LATTER-DAY ITALIAN DRAGONS OF FORLI AND GORO

During 1969 and 1970, reports of a very fierce 16-ft-long lizard-like beast with scales and searing breath emerged from the environs of Forli, a city in Emilia-Romagna, Italy. One eyewitness, 48-year-old Antonio Samorani, even claimed to have been chased for over 200 yards by this creature, and could feel its hot breath. When sceptical police arrived, they were startled to see a series of large footprints in a woodland glade close to where Samorani had allegedly spied the mystery reptile. Nor was this an isolated outbreak of modern-day dragon reports.

In June 1975, a black 10-ft-long reptile of undetermined identity, but described by eyewitnesses as resembling a snake with legs and as thick in girth as a dog, was reported in Goro, just across the Po River

from Venice and again located in Emilia-Romagna. One farmer, Maurizio Trombini, saw it step out of his tomato plants, and he lost no time in fetching the police. Unfortunately, it had gone by the time that they arrived, but like its Forli counterpart it had thoughtfully left behind some discernable tracks, each of which was some 8 in wide.

CROWING CRESTED MYSTERY SNAKES —EXPLAINING THE BASILISK AND COCKATRICE?

In modern times, there have been several eyewitness reports emanating from isolated regions of verdant vegetation amid the otherwise arid desert zones of Morocco and Tunisia that tell of very large snakes bearing a crest or long 'hair' on their head. Such reports readily recall the legendary basilisk, in terms of both location and these snakes' morphology, so could they explain this much-dreaded serpent dragon? Assuming that these snakes are themselves real, it has been suggested that they may constitute relict populations of pythons, analogous to isolated populations of desert-dwelling crocodiles, and that their supposed crests or hair are merely segments of incompletely-shed skin.

Crested mystery snakes are even more commonly reported in tropical Africa, where they have a wide variety of local names, of which the most familiar is the inkhomi ('killer'), but in English parlance they are generally referred to as crowing crested cobras. They are said to sport a bright-red rooster-like coxcomb (but pointing forwards rather than back) and facial

Equally noteworthy is a published report by John Knott from September 1962 in which he recalled how, driving home one evening in late May 1959 from Binga, in the Kariba area of what was then Southern Rhodesia (now Zimbabwe), he inadvertently ran over a large, jet-black snake roughly 6 ft long, mortally wounding it. Knott cautiously stepped out of his vehicle to take a closer look at the snake, and was amazed to see that it bore a distinct crest upon its head, perfectly symmetrical in shape, and capable of being erected by way of five internal prop-like structures. This certainly does not sound like a piece of unshed skin but rather like a true crest, yet no known species of snake possesses one.

Remarkably, a very similar but somewhat shorter mystery snake, complete with coxcomb, wattles, and crowing ability, has been reported in modern times, and by native and Western observers alike, on the West Indian islands of Jamaica and Hispaniola. If such snakes as these on record from Africa and the Caribbean are genuine, they may well have assisted in inspiring the cockatrice legend. There are also reports of crowing (albeit uncrested) snakes from Samoa and Palau in the Pacific.

Back in Africa, certain freshwater mystery beasts, such as the lau reported from the Upper Nile and the lukwata from Lake Victoria, have again been likened by eyewitnesses to lengthy snakes with crests. However, some cryptozoologists have suggested that these aquatic anomalies may be extra-large catfishes, with their 'crests' actually constituting the whisker-like barbels

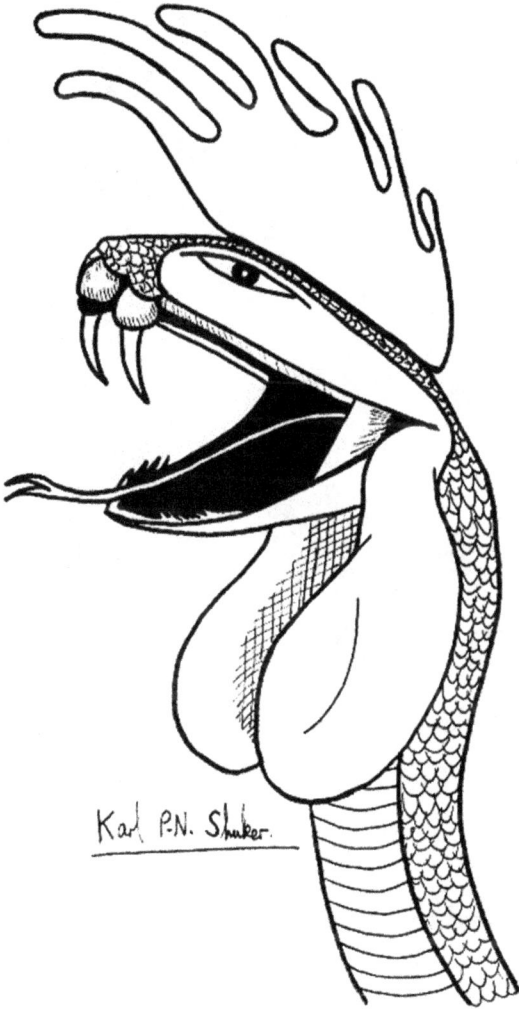

The crowing crested cobra, based upon eyewitness descriptions. (Dr. Karl Shuker)

wattles too, and to crow just like a rooster as well—all of which is very reminiscent of the basilisk's transformed equivalent, the cockatrice. In 1944, Dr J. Shircore from Malawi published a detailed description of what he believed to be the fleshy coxcomb and part of the neck from one of these snakes, but the current whereabouts of this potentially-valuable specimen is not known.

Painting of the mokele-mbembe challenging a hippopotamus. (William Rebsamen)

around the mouth that characterise many catfishes.

THE MOKELE-MBEMBE—A CONGOLESE WATER DRAGON, OR A LIVING DINOSAUR?

According to the resident pygmies here, the vast, remote, and virtually inaccessible Likouala swamplands of the People's Republic of the Congo (formerly the French Congo) are home to a 30-ft-long amphibious 'water dragon' known as the mokele-mbembe ('one that capsizes boats'). They describe it as reddish-brown in colour, with a small head but very long neck and tail, an elephantine body, four sturdy limbs that leave behind large three-clawed footprints, and an appetite for the *Landolphia* gourds, which it browses upon like a reptilian giraffe, sometimes while still partly submerged in water. It is most frequently sighted by the native people in or near a very large body of freshwater known as Lake Tele.

The mokele-mbembe has been sought unsuccessfully by a number of Western expeditions since the 1980s, its most famous and tenacious seekers being now-retired Chicago University biochemist and spare-time cryptozoologist Prof. Roy Mackal, and

Scottish field cryptozoologist Bill Gibbons (who has also sought similar beasts in Cameroon). Of particular interest is that when various animal images have been shown to the natives to gain more information concerning the mokele-mbembe's appearance in the hope of identifying it, the images consistently claimed by alleged eyewitnesses to be closest to it have been ones depicting sauropod dinosaurs from prehistory, such as *Diplodocus* and *Apatosaurus* (formerly called *Brontosaurus*).

Indeed, Mackal and Gibbons both consider a species of living, modern-day sauropod, surviving undisturbed in the Likouala's secluded and relatively inaccessible, inhospitable terrain, to be a plausible identity for this elusive creature. Another popular suggestion is that it may be a very large monitor lizard with an exceptionally long, elongated neck, but no such form of monitor is known from either the present day or the fossil record. And whereas monitors are almost exclusively carnivorous, the mokele-mbembe is entirely herbivorous.

Perhaps the most intriguing aspect concerning the mokele-mbembe is how closely it resembles the Babylonian mushussu dragon as depicted on the Ishtar Gate. And it is known that early glazed bricks identical to those used in this gate's creation have been recorded from Central Africa—leading some naturalists to suggest that perhaps the early Babylonians visited this African region to obtain bricks for the gate, saw mokele-mbembes there, and upon their return to Mesopotamia their descriptions of this creature inspired the depictions of the mushussu on the gate. Boldest but most fascinating of all is the suggestion that at least one young mokele-mbembe may even have been transported back to Mesopo-tamia alive, possibly explaining the so-called dragon that was worshipped in a temple by the Babylonians until killed by Daniel.

DRAGON-BACKED MONSTERS OF THE CONGO AND KENYA

A recurring feature when describing wingless classical dragons is their serrated or spiny back, a feature rarely possessed by real-life reptiles. In the People's Republic of the Congo, however, the mokele-mbembe

The mbielu-mbielu-mbielu. (David Miller under the instruction of Prof. Roy Mackal)

The nguma-monene. (David Miller under the instruction of Prof. Roy Mackal)

apparently shares its Likouala swampland domain with two equally mysterious reptiles that both possess this dragonesque characteristic. One of these cryptids is the triple-named mbielu-mbielu-mbielu. As learned by Prof. Mackal during his two 1980s expeditions here, it is said to be a sturdy quadrupedal beast distinguished by having "planks growing out of its back"— a description irresistibly reminiscent not only of dragons but also of the dinosaur *Stegosaurus*. Yet the mbielu-mbielu-mbielu is reputedly semi-aquatic, whereas stegosaurian dinosaurs were not.

Equally intriguing is the nguma-monene, another amphibious form, equally adept moving in water and on land, and described by the natives as being an extremely elongate and low-slung reptilian beast, almost like a four-limbed snake and measuring up to 30 ft in total length, but with a saw-like dorsal ridge or frill of small triangular projections running down the total extent of its back and its lengthy tail. It also has a forked tongue. This could well be an unknown, primitive species of anguinine-bodied semi-aquatic monitor lizard.

Another dragon-backed mystery reptile is on record from Kenya. Driving through its Great Rift Valley en route to Nairobi one day in summer 1961, missionary Cal Bombay and his wife had to brake sharply in order to avoid colliding with an extraordinary 10-ft-long creature sunning itself in the middle of the road. Dark grey, with a snake-like head and four stubby legs, this reptile's most eye-catching feature was the series of diamond-shaped serrations running from the back of its head down the entire length of its back and tail.

The Bombays had plenty of time to observe the creature closely, because it continued sunbathing for around 20 minutes before finally standing up and wandering off into the bush. Kenyan natives later revealed that this reptile was referred to by them as the muhuru, but there is no living species known to the scientific world that matches its description.

The Bombays' encounter with a muhuru. (William Rebsamen)

THE KATANGA MEGA-SNAKE

It is highly likely that exaggerated observations of extra-large snakes inspired many legends and folktales of serpent dragons, but some veritable giants among snakes that are every bit as monstrous as their dragon counterparts may truly exist too. For instance, in 1959, while flying over some termite mounds and vegetation in Katanga (now renamed Shaba), a province of the Democratic Republic of Congo, highly-experienced Belgian aviator Colonel Remy van Lierde spied an immense python-like serpent emerge from a hole in the ground and rear upwards, directly towards his helicopter!

Despite his great alarm, van Lierde was able to photograph this terrifying reptile, and careful analysis of the background topography in the image enabled the snake's total length to be accurately estimated. Incredibly, it had apparently measured at least 60 ft. This is not an isolated incident either. The local tribes are so familiar with mega-pythons like this one that they even have their own specific name for them—the pumina.

THE KONGAMATO—A LIVING PTEROSAUR AND A MODERN-DAY DRAGON?

In medieval times, fossil remains of those famous flying reptiles from prehistory known as the pterosaurs were often deemed to be the earthly bones of winged dragons. In an intriguing cryptozoological twist on this subject, however, modern-day native reports of a small but greatly-feared winged dragon lurking within the fever-ridden, scarcely-penetrable Jiundu swamps

Modern-day reconstruction
of *Rhamphorhynchus*. (Wikipedia)

of Zambia may, in the opinion of some investigators, refer to a surviving species of pterosaur.

According to native testimony, the kongamato has a pair of leathery wings, a long tail, and a beak brimming with teeth—a description irresistibly reminiscent of a small type of pterosaur known as *Rhamphorhynchus*. Due to their inhospitable nature, let alone the presence of the kongamato within their febrile interior, the Jiundu swamps are entered only by the bravest tribesmen but according to the testimony of those who have done so, this dreaded creature is most commonly encountered near lakes and rivers, and is said to prey upon fishes, skimming across the water's surface and impaling them upon the sharp teeth of its open jaws.

During the early years of the 20th Century, the kongamato was described by local hunters to one Westerner, Frank H. Melland, as a small flying crocodile with black bat-like wings. Could one ask for a better description of a rhamphorhynchoid pterosaur—or a modern-day dragon?

Other pterosaurian cryptids, including a small version known as the ropen and a far bigger form called the duah (said to boast a wingspan of 20 ft and a glowing underside), have also been reported in recent years from New Guinea and certain offlying islands. They are said to be carrion feeders, attacking funeral processions and even robbing freshly-dug graves to devour human corpses.

On 8 August 1981, a pair of winged cryptids that almost collided with a car on Tuscarora Mountain in Pennsylvania, USA, were each soaring on huge unfeathered membranous wings like those of bats, possessed a straight beak, and bore a long bony backward-pointing projection on top of its head. This description closely corresponds with one of North America's most familiar prehistoric pterosaurs—*Pteranodon*. Indeed, some eyewitnesses of flying mystery beasts spied elsewhere in the USA have positively identified them as *Pteranodon*.

THE ARTRELLIA—PROVING THE PAPUAN DRAGON

During the 'Operation Drake' expedition to Papua New Guinea in 1980, led by explorer Colonel John Blashford-Snell, the team heard many stories of a gigantic dragon-like creature called the artrellia, said to measure up to 30 ft long, and allegedly stalking the rainforests there. One member of the team, eminent British conservationist Ian Redmond, heard the heavy footfalls and even briefly spied the head of an exceptionally large lizard that could have been one of these 'dragons'.

Just for once, however, a mystery beast was conclusively identified when a small 6-ft-long artrellia was actually captured alive. It proved to be a juvenile specimen of Salvadori's monitor lizard *Varanus salvadorii*, which is known to attain a total length of up to 15 ft. Perhaps even bigger specimens of this species still await detection, however, because such creatures would clearly explain the reports of dragons here. Also presently unconfirmed in New Guinea are reports of giant amphibious monitors, known locally as the au angi-angi.

Enormous lizards have been reported from mainland Australia too, especially scrub-covered mountain ranges in New South Wales, around the Murray River, and further south near Lake Alexandrina in Victoria, and are known to the aboriginal nations as the mungoon-galli. Cryptozoologists have speculated that perhaps these are relict survivors of the officially extinct giant monitor *Megalania prisca*. This very formidable member of Australia's all-but-vanished megafauna is believed by some palaeontologists to have attained a total length of 23 ft and to have weighed as much as 700 lb, but it died out around 40,000 years ago—didn't it?

Giant freshwater lizards that may again be monitors but said in some instances to measure 20 ft or more in total length have lately been reported from the Gir region of India, where they are called the jhoor, as well as from neighbouring Asian countries. Indeed, one such beast was sighted in a northern lake in Bhutan by no less eminent an eyewitness than this country's reigning monarch! Monstrous lizards like these could readily have encouraged Oriental lore concerning huge wingless dragons.

Encounter with a mungoon-galli.
(William Rebsamen)

THE GOLDEN DRAGONS OF TASEK BERA
Tasek Bera is a very large, deep lake in the Malaysian state of Pahang, and according to the traditions of the local Semelai people it was once home to a number of huge water dragons whose scales were slate-grey when young but became golden as they matured. They had very long necks, serpentine heads bearing a pair of snail-like horns, sturdy bodies, and long tails. As they never emerged onto land, however, no-one had ever seen their limbs. These beasts' presence was confirmed by their loud, trumpeting cry. As documented in his book, *The Lost World of the East* (1958), explorer Stuart Wavell paid two visits here during

the 1950s to search for them, but he never saw one.

However, on one occasion he did hear a strident, twice-uttered staccato sound emanating from the centre of the lake that matched the native description of these creatures' cry. Was it just the trumpeting of an elephant—or the voice of a golden dragon?

Reconstruction of the Mongolian death worm's appearance based upon local descriptions. (Ivan Mackerle)

THE MONGOLIAN DEATH WORM

One of the most extraordinary serpentine mystery beasts on record must surely be the allghoi khorkhoi ('intestine worm') of the Gobi Desert, more commonly referred to as the Mongolian death worm. Around 5 ft long, it earns its native name from its appearance, which, according to the nomadic people inhabiting its sandy domain, greatly resembles a blotched, scarlet-hued, animate intestine! It has no recognisable head or tail, nor any identifiable sense organs, and for much of the year it remains hidden beneath the desert sands. However, it can sometimes be seen resting on the surface during June and July, which are the two hottest months in this vast arid wilderness.

The death worm earns its Western name from its two very different (but equally effective) alleged modes of dealing out death to those who trouble it. It can raise one end of its body upwards and accurately squirt a stream of highly acidic and extremely venomous liquid at any would-be aggressor. And if anyone, or anything (such as a camel), should inadvertently tread upon a death worm concealed beneath the sand, they will drop down dead instantly. There is even one case on record of a visiting geologist who accidentally prodded a hidden death worm with an iron rod, only to collapse onto the ground, dead, in front of his horrified colleagues.

What makes this incident so alarming but also so remarkable is that the only way the creature could have killed the geologist when he merely touched it indirectly, using the metal rod, is by electrocution. Yet although several different types of fish that generate electricity to stun their prey are documented, no terrestrial animal is known to possess a comparably potent ability.

Needless to say, if the death worm is a genuine creature, currently unrecognised by science but well known to (and greatly feared by) the local people, early reports of this lethal entity may well have helped shape the legends of small yet deadly desert-dwelling serpent dragons such as the basilisk and the amphisbaena.

Reconstruction of a living Nepal dragon. (William Rebsamen)

THE 'LIVING TREE TRUNK' DRAGONS OF NEPAL

Among the strangest reptilian cryptids on record from modern times are the so-called dragons of Nepal's secluded jungle valleys. According to local testimony, as well as that of longstanding Nepalese missionary Reverend Resham Poudal (who saw one at close range), these bizarre creatures are approximately 36 ft long, limbless, and greatly resemble fallen tree trunks in form, heightened by their preferred mode of prey-capturing behaviour. This involves lying motionless with their huge crocodile-like jaws wide open, and waiting until a suitable prey victim approaches close enough to be seized and devoured.

Such creatures readily recall descriptions of mythical serpent dragons—huge and limbless but with very long crocodilian jaws rather than the shorter, more slender jaws of true snakes. And if they are real, they must certainly have inspired local belief in dragons.

THE TIBETAN YAK-SNATCHER OF LAKE WENBU

Lake Wenbu (aka Menbu) is a very large and fish-rich body of freshwater in Tibet. During the 1980s, this hitherto little-known lake attracted media headlines worldwide, due to various reports that it harboured a long-necked water dragon of murderous intent, which had not only snatched a local herder's yak that had been left to graze on the lake's shores but was also being blamed for the unexplained disappearance of a farmer who had been rowing on the lake. Nowadays, however, Lake Wenbu has once more sunk into obscurity, as no further news—nor, indeed, any dragon—has since emerged.

THE SUCURIJU GIGANTE— A DRAGON-SIZED ANACONDA?

According to the record books, the common or green anaconda *Eunectes murinus* of South America rarely exceeds 20.5 ft. However, many reports have emerged from various tributaries of the mighty Amazon River in Brazil appertaining to a monstrous form of anaconda known as the sucuriju gigante, whose size is supposedly far in excess of any specimens formally measured and documented by science. One of the most famous, or infamous, examples was a stupendous individual shot in 1907 by the subsequently-lost explorer Lieutenant-Colonel Percy Fawcett as it began to emerge from the Rio Abuna. Part of the snake's huge body lay on the riverbank, and this measured 45 ft; the remainder was still in the water, and Fawcett estimated this portion

at 17 ft—thus yielding a total length of 62 ft, a size that any serpent dragon would have been proud to have attained.

Representation of the minhocão.
(Lance Bradshaw)

MINHOCÃO—A MODERN-DAY SERPENT DRAGON WITH HORNS?

On record from Brazil and Argentina are many reports of an enormous vermiform creature, 30 ft or more long and bearing a small pair of horns on its head, which excavates such enormous underground burrows that it can divert the course of rivers and cause wholesale collapse of terrain directly above it. This subterranean terror is known as the minhocão, and bears more than a passing resemblance to traditional folklore accounts of serpent dragons from this region of South America.

If such a devastating creature truly exists, the most plausible identity for it is a giant species of caecilian—a remarkably earthworm-like type of limbless amphibian, some species of which bear cephalic tentacles that do look a little like horns, and spend much of their lives burrowing underground. Having said that, no known species of caecilian even remotely approaches the size claimed for the minhocão, but allowing for exaggeration caused by shock at seeing such a beast if it briefly emerges while digging (as affirmed by eyewitnesses), an unknown caecilian species of less dramatic dimensions may well be sufficient to explain this cryptid.

AN IMPERVIOUS DRAGON
IN THE MATO GROSSO?

In or around 1933, while travelling on the River Marmore in Brazil's Mato Grosso, Swedish traveller Harald Westin spied an astonishing dragon-like beast walking along one of the river's banks. Grey in colour and approximately 20 ft long, it resembled a distended boa constrictor or python, but sported a very large alligator-like head, lizard-like feet, and fairly small red eyes. Westin took aim with his gun and shot this daunting, potentially dangerous creature of wholly unknown identity, but the bullet did not seem to harm it at all, as it merely continued on its way, lumbering off towards the river.

SEA SERPENTS AND LAKE MONSTERS
—THE ORIGINAL WATER DRAGONS?

For centuries, sea-going voyagers have reported sightings of very large, mysterious marine beasts of many different forms that remain unexplained even today by mainstream zoology but which are usually referred to by the umbrella term 'sea serpent', even though it is clear that more than one animal species is involved (and, equally, that none are true snakes!). In 1968, French cryptozoologist Dr. Bernard Heuvelmans published his seminal book *In the Wake of the Sea-Serpents*, in which he distinguished several entirely separate categories of mystery beast collectively responsible for sea serpent sightings. Not all of these categories are still recognised as valid today by other cryptozoologists, but of those that are, at least three are relevant to the subject of whether observations of sea serpents have contributed to the development of water dragon folklore and mythology.

The longneck category of sea serpent. (Tim Morris)

One of these sea serpent categories is termed the longneck, due to this cryptid's most distinctive feature—a very long, vertically-held neck that resembles a periscope. Based upon no less than 48 independent eyewitness accounts collated, assessed, and accepted as valid by

Heuvelmans, the longneck is apparently 15-60 ft long, with two pairs of flippers, a phenomenally fast swimming ability, and a worldwide distribution (but particularly common off North America and Scandinavia). Heuvelmans considered it most likely to be a highly-specialised, giant species of long-necked seal. However, some longneck sightings also report a long tail for this sea serpent, which seals do not have. Consequently, a number of cryptozoologists have rejected a seal identity for the longneck in favour of an evolved plesiosaur.

Plesiosaurs were famously long-necked prehistoric reptiles that also possessed distinct tails but became extinct at least 65 million years ago and, based upon current belief, had necks that were held horizontally, not vertically. If, however, a plesiosaur lineage had survived up to the present day, 65 million years of continued evolution may have engineered a much-modified plesiosaur, one whose neck could possibly have become much more flexible, enabling it to be held vertically.

Such a creature would correspond very closely with the longneck sea serpent. It would also recall the sea-worm (sea dragon) ornately portrayed as protective, 'all-seeing' figureheads on the prows of many Viking ships dating back a thousand years or more. One of the most famous was the sea dragon prow from the Oseberg ship burial of c.825 AD. Did the maritime

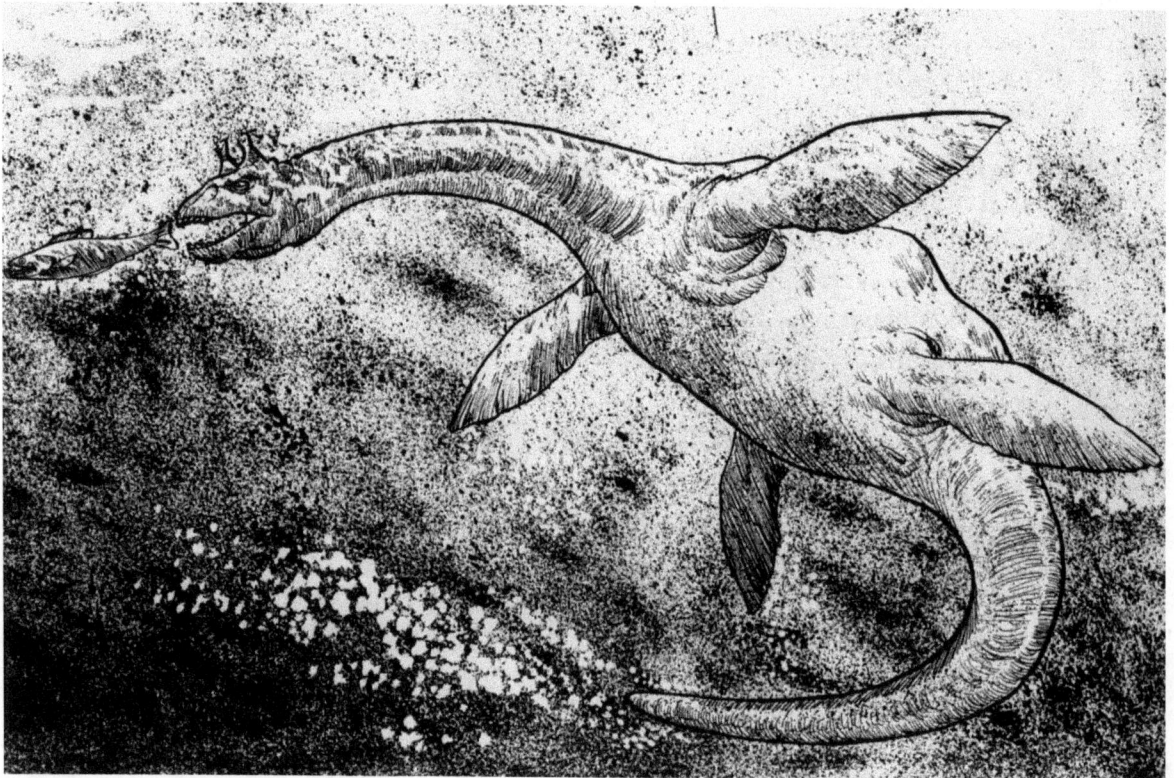

The Loch Ness monster. (Richard Svensson)

Vikings base their vessels' figureheads upon direct sightings of a real but currently-unrecognised species of long-necked sea serpent?

Interestingly, aquatic cryptids very similar in form to the longneck are also on record from many temperate freshwater lakes around the world, most famously Loch Ness in Scotland and Lake Champlain in the USA. These too are considered by some to be surviving evolved plesiosaurs, and by others to be long-necked seals.

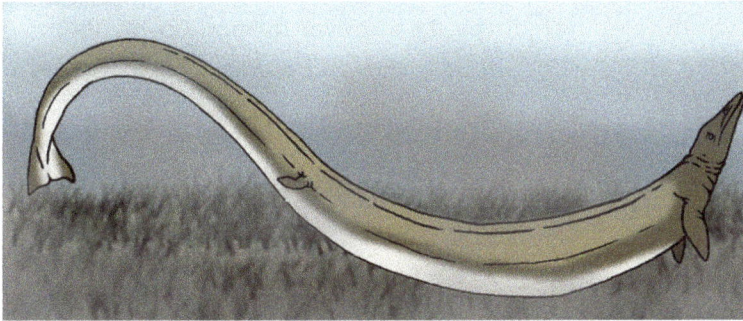

A reconstruction of *Basilosaurus*. (Tim Morris)

Another of Heuvelmans's categories of sea serpent with potential dragon connections is the many-humped or serpentiform sea serpent, reported from seas all around the world but especially the waters off New England, USA. As its names suggest, this creature is allegedly extremely elongate and snake-like in general shape, and very long too (estimated at 60-100 ft). However, it undulates its body not horizontally (like snakes and other reptiles do), but vertically (like mammals do), throwing its body into a series of buoy-like humps, thereby indicating that it is a mammal. This description readily recalls the numerous marine serpent dragons in folklore and legends from many cultures globally.

The most popular identity for the many-humped sea serpent among cryptozoologists is a surviving evolved archaeocete. These were—or are?—extremely lengthy and remarkably elongate whales (as exemplified by *Basilisaurus*), which officially died out around 30 million years ago but which, if lingering on today and sufficiently evolved to have become more flexible than their antecedents, would fit the description of the many-humped sea serpent very satisfactorily.

There are many reports of cryptids resembling the many-humps from freshwater localities too, in particular Canada's Lake Okanagan (whose serpentiform monster is known as Ogopogo), and a number of Irish lakes or loughs (where the creatures are termed horse-eels).

Much rarer in terms of sightings than the longneck and the many-humps is a third dragon-related category of sea serpent proposed by Heuvelmans—the marine saurian. In general appearance, it resembles a huge crocodile, but with flippers instead of normal claw-footed legs. Millions of years ago, marine crocodile-related reptiles with flippers did indeed exist—the thalattosuchians, and a surviving evolved thalattosuchian is a popular identity among cryptozoologists for the marine saurian. Another oft-proffered suggestion is a surviving evolved mosasaur—a huge sea lizard closely related to today's monitor lizards, which again had flippers instead of legs.

Measuring 50-60 ft long, the marine saurian appears to be restricted to tropical seas and oceans, with a number of good modern-day sightings reported from the waters around New Zealand and other Pacific areas. There is no doubt that if this cryptid truly exists, it would very closely match traditional descriptions in myths and legends of sea dragons, and a modern-day mosasaur has been mooted as an explanation for the biblical Leviathan.

DUE TO THE ABUNDANCE of cryptic reptiles documented here, it seems likely that there may well be some veritable dragons still lingering in the more remote corners of the world, currently beyond the reach of persecution from latter-day St. Georges and also from the revealing glare of scientific scrutiny. Long may the former situation continue, but as far as the latter one is concerned, the quicker that such creatures receive official scientific recognition the better. Otherwise there is the very ironic, tragic prospect that they could ultimately become extinct before their reality was ever formally confirmed.

PART II:
DRAGONS IN CULTURE

CHAPTER 5:
THE SYMBOLISM OF DRAGONS

IN TERMS OF physical diversity, the dragon has no equal among fabulous, legendary beasts. The same is also true of its global ubiquity and popularity as a multipotent symbol of profound fundamental significance, whether it be in religion, psychology and dreams, astrology, astronomy, alchemy, heraldry, tattoos, or feng shui.

DRAGONS IN RELIGION
The presence and roles of dragon symbolism in religion exhibit an extraordinary occidental-oriental schism. Expressed in its most basic form, the Western dragon is a symbol of evil and destruction whereas the Eastern dragon is a symbol of good and creation. Western dragons are to be despised and slain, Eastern dragons are to be respected and venerated.

The earliest dragons, arising in Babylonian and other ancient Middle Eastern mythology, symbolised both the earthly rulers of these lands and also their divine deities, whereas in China, Japan, and elsewhere in the Far East, as airborne entities, they came to represent the very sky itself, as well as all matter, life, and the life-giving waters of rain, lakes, rivers, and the seas. As such, Oriental dragons were venerated as gods, generally benevolent and very wise, but quite capricious too, especially if not accorded sufficient reverence.

Dragon worship and symbolism in the East is long-established. The earliest known Chinese dragon statue was discovered in Henan during 1987, measures almost 6 ft long, and dates back to the fifth millennium BC from the Yangshao culture. According to traditional Chinese custom, there are four principal dragon kings, which are dragon-headed humanoid deities all demanding of veneration. Collectively, they rule the waters in all four cardinal directions—the North (applying to Lake Baikal),

The author alongside Oriental dragons carved upon pillars at
the Thean Hou Temple in Kuala Lumpur, Malaysia. (Dr. Karl Shuker)

the East (East China Sea), the South (South China Sea), and the West (Indian Ocean and the waters beyond it). Even today, dragon king worship continues in many areas of China, and remains a very popular, recurrent theme in Chinese New Year celebrations.

In Tibetan Buddhism, the dragon is one of four greatly-revered, spiritually-endowed creatures. It is specifically identified with power, strength, and dominance, but also compassion, as well as all aspects appertaining to water.

Even in the West, there are certain faiths and traditions that cast the dragon as a benign beast. In Celtic traditions, for example, the dragon symbolises fertility, wisdom, and immortality—a far cry from its malign status elsewhere in Europe. Indeed, in Druidism, dragon symbolism and veneration occupy significant roles, and feature more than one type of dragon.

Philip and Stephanie Carr-Gomm have been respectively Chief and Scribe of the Order of Bards, Ovates and Druids—one of the largest international Druid groups. In their book *The Druid Animal Oracle* (1994), which is linked to a set of 33 sacred animal-interpretation cards, they elucidate four elemental dragon types and their symbolism in the Druid religion.

The water dragon or draig-uisge symbolises passion, depth, and connection, conjuring forth into the light of day all that is concealed, such as distant memories and long-repressed desires, enabling a person to confront and deal with them, assimilating their power into their consciousness to

A spectacular Oriental dragon carved in the grand staircase, Thean Hou Temple, Kuala Lumpur. (Dr. Karl Shuker)

achieve a greater depth of soul as well as much-needed balance and stability. The earth dragon or draig-talamh symbolises power, potential, and riches, enabling a person to recognise their own potential, a veritable treasure trove of powers and capabilities that can be harnessed and utilised. The air dragon or draig-athar symbolises inspiration, insight, and vitality, manifesting as a sudden lightning-like flash or bolt of illumination and clarity to a person's thoughts and imagination. And the fire dragon or draig-teine symbolises transmutation, mastery, and energy, unleashing vitality, enthusiasm, and courage as well as enhancing a person's ability to overcome obstacles, and find the necessary energy to deal with life's problems.

St. George slaying a Western dragon,
symbolising evil and paganism,
by Albrecht Altdorfer, 1511.

Elsewhere in Western traditions, conversely, the dragon represents evil and darkness, which must be overcome to achieve supremacy of the land and for goodness and light to triumph. In Christianity, the dragon is particularly allegorical, identified directly with the devil and also more generally with sin, heresy, and paganism. Consequently, medieval European myths and legends of knights and other heroes battling dragons are representations of God or Christ battling Satan and his minions, or the Church fighting the good fight against sinners, heretics, and pagans. This is why such conflicts were so popular a subject for European religious art—paintings, church architecture, stained-glass windows. Indeed, as far back as the 12th-Century work *Physiologus* (translated in modern times by T. H. White as *The Book of Beasts*), the dragon was fully recognised as a major source of medieval religious and secular iconography.

In the Christian tradition, one specific dragon type, the basilisk, has come to symbolise the antichrist. In more general, medieval symbolism it represented lust, deceit, and disease (most notably syphilis). Yet, paradoxically, its image was also used to ward off evil.

DRAGONS IN PSYCHOLOGY AND DREAMS

In psychology, the dragon is an archetype—a term utilised by Swiss psychiatrist Carl Jung (1875-1961) to denote a fundamental, primordial symbol or prototype that the human brain commonly utilises in order to convey subconscious messages. Archetypes constitute the content of the collective unconscious (another Jung-associated term, denoting the part of a person's unconscious that is derived from the experiences of their race or species). As an archetype, the dragon represents primal chaos that must be overcome in order for a person to gain peace of mind. It can also symbolise bestial desires that need to be confronted and

Woodcut of St. George and the dragon, from
The Life of Saint George (1515), Alexander Barclay.

conquered, as well as embodying the spirit of change, and is the personification of water too.

Psychologically, the image of St. George fighting the dragon represents the struggle of a person's conscious ego (conveyed by St. George) to escape from the all-encompassing unconscious (the dragon) and gain control of it. In addition, slaying the dragon is considered to be a psychological metaphor for breaking those ties with a person's mother that are detrimental to his finding his own psychic individuality.

In addition, the dragon can be identified with Mother Nature's generative powers, and thus a womb. It can also symbolise a person's own sexuality if it frightens him, but in such circumstances the dragon should not be slain, but rather tamed.

In view of the above, it will come as no surprise to discover that a symbol as potent psychologically as the dragon often features in dreams, but how are its presence, appearance, and behaviour in them interpreted? In other words, if you dream about a dragon, what does it mean?

Materialistically, it traditionally signifies that the dreamer may well be coming into riches, treasure, or very good fortune, especially if the dream dragon is guarding treasure. It can also portend a forthcoming encounter with an important person, a figure of great authority and power, who may enhance the dreamer's future in some significant manner. At a deeper level of interpretation, the dreamer himself is the treasure and the dragon is a protector,

someone destined to enter his life and provide for him a new level of security.

Psychologically, however, a dragon's presence in a dream indicates that the dreamer is content to be controlled by fiery emotions, sexual passions and/or irrational beliefs, to the extent that enemies may take advantage of this weakness and use it against the dreamer. A dragon should thus be seen as a warning to the dreamer to take control of the wilder, baser side of his nature, and curb its excesses in order to protect his future.

Dreaming about being physically attacked by a dragon indicates that the dreamer's passions are out of control. It also symbolises his fear of the unknown, or, more specifically, the fear of discovering what lies locked inside his own unconscious mind.

Battling a dragon suggests that the dreamer has to overcome some fear, and also to reconcile opposing forces or passions within himself, in order to achieve self-recognition and gain self-esteem.

Briefly catching sight of a dragon in a dream but without making any form of contact with it can represent illusive happiness—happiness that will prove to be as intangible as the dream heralding its coming.

Different types of dragon mean different things too in dreams. A classical dragon with wings, for instance, can epitomise a transition, an ascent from a lower to a higher level of maturity. A many-headed hydra, conversely, signifies that the dreamer is plagued by a recurrent problem, one that he has tried to deal with several

Dreaming of a hydra denotes the persistence of a recurrent problem.

Persons born in the Year of the Dragon are believed to possess the positive traits of artistry, decisiveness, dignity, directness, eccentricity, empathy, fieriness, generosity, intellect, loyalty, magnanimity, nobility, passion, pioneering ability, pride in worthy achievements, self-assurance, stateliness, strength, and vigour. However, they also display the negative traits of arrogance, brashness; a demanding nature, dogmatic attitude, imperiousness, impetuosity, intolerance, rebellion, tactlessness, tyranny, and violence.

times but always unsuccessfully, so it is still appearing in his life, awaiting a satisfactory, conclusive resolution. And whereas a sea dragon's appearance in a dream does not normally bode well for the dreamer, if he observes it out of the sea, on land where it is rendered less adept or helpless, then this is a good omen.

In Western astrology, when plotting a person's chart via the most common form of zodiac, known as the Tropical Zodiac, the

DRAGONS IN ASTROLOGY

In Chinese and Mongolian astrology, there are twelve signs of the zodiac, all of which are animals. Of these, one—the fifth sign—is a mythological creature, the dragon. The dragon sign is linked with the polarity or direction of yang, the birthstone of bloodstone, the colours of red and violet, the fixed element of wood, the compass direction of East-south-east, the ruling hours of 7 am to 10 am, the season of spring, the month of April, and the motto "I Reign."

The dragon plays an integral role in Chinese astrology.

astrologer begins 0 degrees Aries (i.e. the First Point of Aries, the zodiac's first sign) at the vernal (spring) equinox. However, there is another, less familiar zodiac version that is named after the dragon, and known therefore as the Draconic Zodiac. In this version, 0 degrees Aries is begun at the moon's North (Ascending) Node, which is also termed the Caput Draconis—'Dragon's Head.' This means that whereas the planets all remain in the same houses and have the same aspects (angular relationships) to each other as they do in the Tropical Zodiac, the signs and degrees are altered. Astrologer Pamela Crane popularised the Draconic Zodiac in her book *Draconic Astrology* (1987), claiming that this zodiac gives an insight into a person's higher spiritual purpose, supplementing the Tropical Zodiac chart.

DRAGONS IN ASTRONOMY

The International Astronomical Union (IAU) recognises 88 different constellations of stars in the night sky. No less than seven of them have been named, for various reasons, after three famous dragons from Greek legends and four notable persons featuring in those same legends.

Draco is one of the original 48 constellations first formulated as long ago as the 2nd Century AD by Egypt-based Graeco-Roman scholar Claudius Ptolemy in his mathematical and astronomical treatise the *Almagest*, and is named directly after the dragon ('Draco' is Latin for 'dragon'). It is perpetually visible in the northern hemisphere, where it is the eighth brightest constellation in the entire night sky.

Due to the distribution of its stars, Draco is usually represented as a very elongate, coiling serpent dragon, and some astronomers believe that Ptolemy's naming of it may have been inspired by a Roman myth in which one such dragon called Draco was killed by Minerva, goddess of wisdom, who then hurled its dead body up into the heavens. Others, conversely, prefer to identify it with Ladon, the many-headed serpent dragon of Greek mythology that guarded the golden apples of the Hesperides until defeated by Heracles (aka Hercules). Moreover, in the night sky, one of the constellations directly adjacent to Draco is none other than Hercules, with one of his feet set resolutely upon Draco's head.

Worth noting, however, is that until around the 6th Century BC, the constellation now called Draco was represented as a winged dragon. Then its wings were 'removed' by astronomers and converted into what became the constellation of Ursa Minor (the Little Bear), in order to assist navigation at sea by sailors.

Another of Heracles's dragon foes is also represented in the night sky by a constellation—the largest constellation of all, in fact. This is Hydra, named after the formidable and near-invincible many-headed lindorm dragon from Lernea, which was eventually slain by Heracles as one of his twelve great labours. It is another of the 48 constellations first recognised by Ptolemy. Yet despite its name, the distribution of its stars across the sky is such that Hydra the constellation bears much more of a resemblance in shape to a writhing single-headed serpent than to the polycephalic monster

Draco and Ursa Minor constellations, in *Urania's Mirror* (a boxed set of 32 constellation cards engraved by Sidney Hall and first published in c.1825).

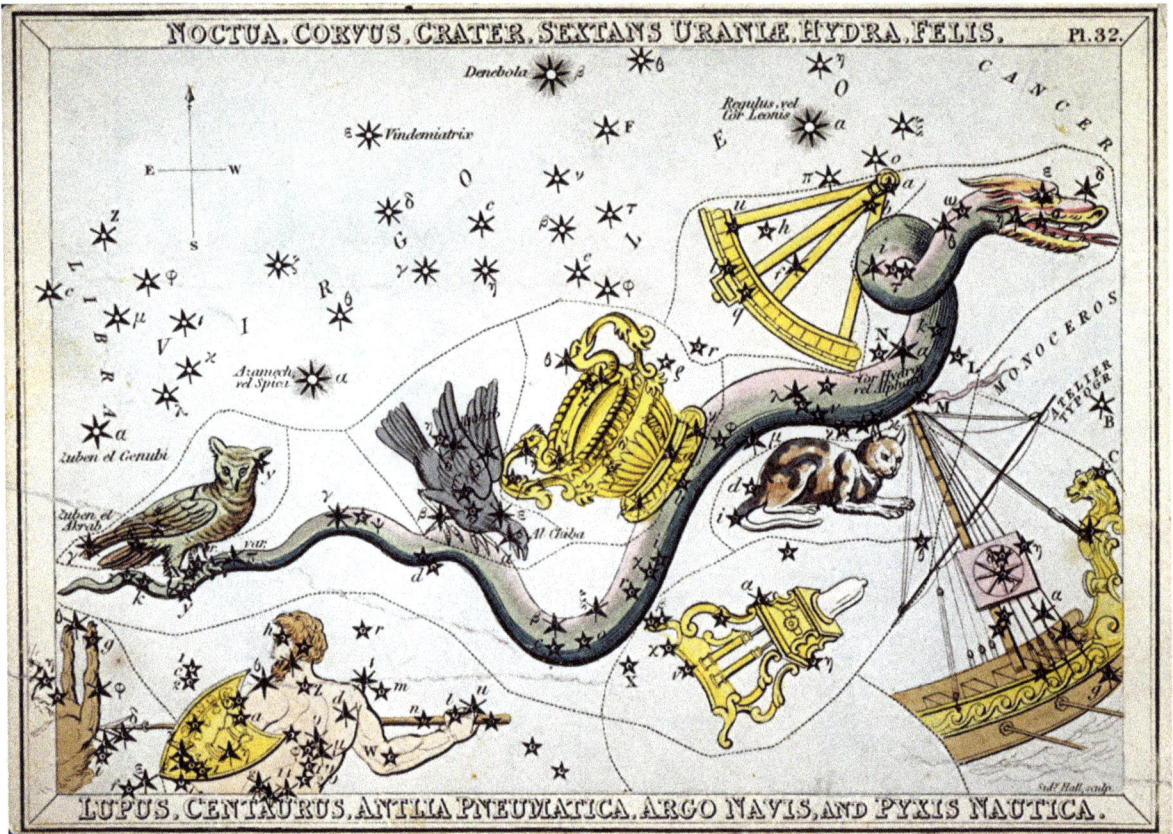

Hydra constellation, in *Urania's Mirror*.

battled by Heracles. This in turn can cause a degree of confusion with another constellation, Hydrus, which is represented as a water snake. Moreover, Hydra itself is adapted from an ancient Babylonian serpent constellation.

As already noted, Heracles is also represented in the night sky by a constellation—Hercules (his Roman name). Fifth largest in the night sky and another of Ptolemy's 48 originals, this constellation, interestingly, is believed by some researchers to have originally been united by ancient Babylonian sky-watchers with Draco, yielding a serpent-bodied human.

Draco and Hydra are two of the three dragons represented in the night sky; the third member of this trio is Cetus. This was the rapacious sea-dwelling serpent dragon from whom the Greek hero Perseus rescued Andromeda—an Ethiopian princess due to be sacrificed to Cetus in order to placate the sea god Poseidon after he had been angered by vain boasts made by her mother, Cassiopeia. The fourth largest constellation, Cetus suitably shares its portion of the night sky with various others representing water-related subjects, including Aquarius (the water carrier), Pisces (the fishes), and Eridanus (the river). Despite its classical origin as a marine dragon, however, Cetus is nowadays more commonly represented astronomically as a whale.

Also represented by constellations in the heavens are three of the major human figures featuring in this Greek legend— Andromeda, Cassiopeia, and Perseus, all three of which were first designated yet again by Ptolemy. The constellation of Andromeda is located in the northern sky, is separated from Cetus merely by Pisces and partly by Aries, and contains within its borders the Andromeda Galaxy, which is the nearest galaxy to our own Milky Way. This constellation is derived from an

equivalent one recognised in Babylonian astronomy that was known as Anunitum, the Lady of the Heavens.

Fittingly, another constellation that lies close to Andromeda in the sky is Cassiopeia, named after the princess's boastful mother. Cassiopeia is also bordered by the constellations representing her husband, Cepheus (king of Ethiopia), and her daughter's rescuer, Perseus.

One of the stars in the constellation of Perseus is Algol, which represents the snaky-haired head of the petrifying gorgon Medusa, another monster slain by Perseus the hero. This constellation may be derived

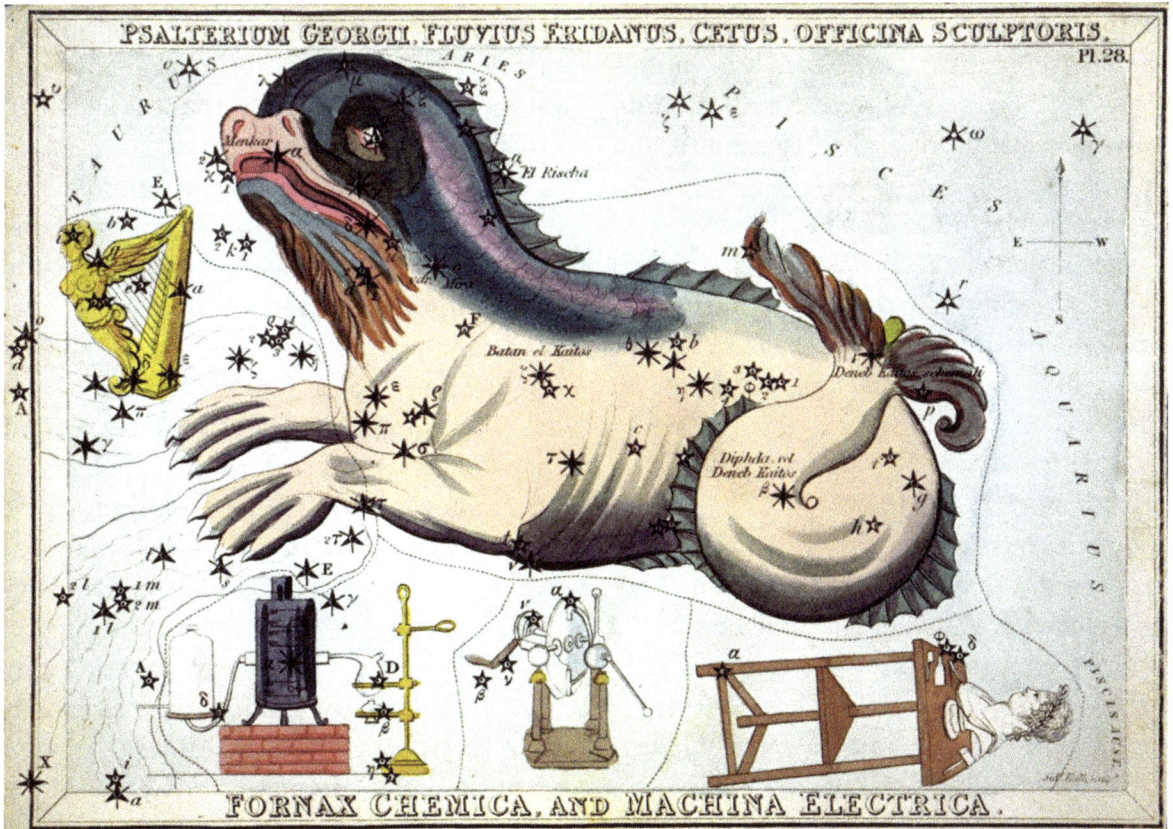

Cetus, whale-dragon constellation, in *Urania's Mirror*.

from an ancient Babylonian predecessor known as the Old Man.

In Chinese astronomy, the stars are grouped in very different constellations from those recognised by the IAU in the Western world. Within the Chinese system, the night sky is divided into 31 different regions, known collectively as The Three Enclosures and The Twenty-Eight Mansions. The stars in the latter group occupy the zodiacal band, and are divided into what are termed the Four Symbols. Each of these symbols represents a different compass point and a different season, and contains seven of the Twenty-Eight Mansions. One of the four symbols is the Azure Dragon of the East and spring (the others being the Black Tortoise of the North and winter, the White Tiger of the West and autumn, and the Vermilion Bird of the South and summer).

According to Chinese legend, the reincarnation of the Azure Dragon's star was General He Suwen of the Liao Dynasty. And in Japanese tradition, the Azure Dragon is a guardian of cities, particularly Kyoto in the east.

DRAGONS IN ALCHEMY
Centuries before chemistry and medicine as modern-day sciences came into being, there was alchemy. Yet whereas some historians nowadays look upon alchemy itself as a protoscience, it was more than that, inasmuch as it also embodied a spiritual, esoteric component. This was heightened in popular perception by virtue of the mystical, cryptic, and sometimes ostensibly misleading manner in which its processes, formulae, and other applications were presented, both verbally and pictorially— which in turn was a vital necessity if its practitioners were to stay alive.

Many rulers and influential church figures across Europe and elsewhere considered alchemy to be diabolical, and so declared it illegal, with those found indulging in this 'black art' summarily put to death, often by very painful, torturous methods. Consequently, treatises on alchemy were to all intent and purposes lavishly-illustrated, heavily-disguised codes. To the layman, such works seemed to be little more than retellings of myths and legends, but they could be readily deciphered by the adept to reveal the arcane alchemical lore concealed within their allegorical, ambivalent words and images.

One of the most popular symbols employed by alchemists within their texts was the dragon, which was portrayed in many different ways and had multiple meanings. It often appeared in illustrations depicting it being fought by knights or by other dragons, thereby giving the outward impression that such works were documenting legendary encounters between dragons and their fabled destroyers. However, the alchemical explanation of such iconography was very different.

For instance, as the dragon in alchemy most frequently symbolised the prima materia (matter in its perfect, unregenerated state), a picture depicting a dragon being slain by a knight was actually an instruction to the alchemist (represented by the knight) to reduce base metals to a non-metallic state via transmutation, with the piercing of the dragon's body by a lance or arrow signifying the penetration of physical matter by alchemical fire. The specific type of dragon portrayed was also of significance.

An alchemical diagram of a knight slaying a dragon, representing the transmutation of base metals, from *The Book of Lambspring* (1599), a famous alchemical work

A wingless classical dragon symbolised sulphur, the fixed (stable) element, and the male sex, whereas a winged classical dragon represented mercury, the volatile (changing) element, and the female sex. An ouroboros or tail-biting dragon specifically symbolised a metallic alloy or amalgam warmed until all traces of its metallic constituents had been rendered invisible, and, more loosely, represented unity in all things. A salamander symbolised rebirth and also nature overcoming nature (the salamander surviving amid the flames of fire); spiritually, it personified those who successfully travelled unblemished through the fires of passion.

Colours also embodied a wealth of hidden meanings, which varied according to the precise form or activity of the dragon so portrayed. For example, in general a green dragon personified the positive universal spirit present in all things, whereas a red dragon personified the negative universal spirit. If a green dragon was specifically portrayed devouring the sun, however, this signified the dissolving of gold in aqua regis—a potent, highly corrosive mixture of nitric and hydrochloric acids. A polycephalic dragon with heads of differing colours represented a sequence of reactions, its heads' respective colours revealing the sequence's exact order.

Two dragons facing in opposite directions from each another represented the attainment of immortality, but if facing one another they symbolised the alchemist's own search for it. A male and female dragon

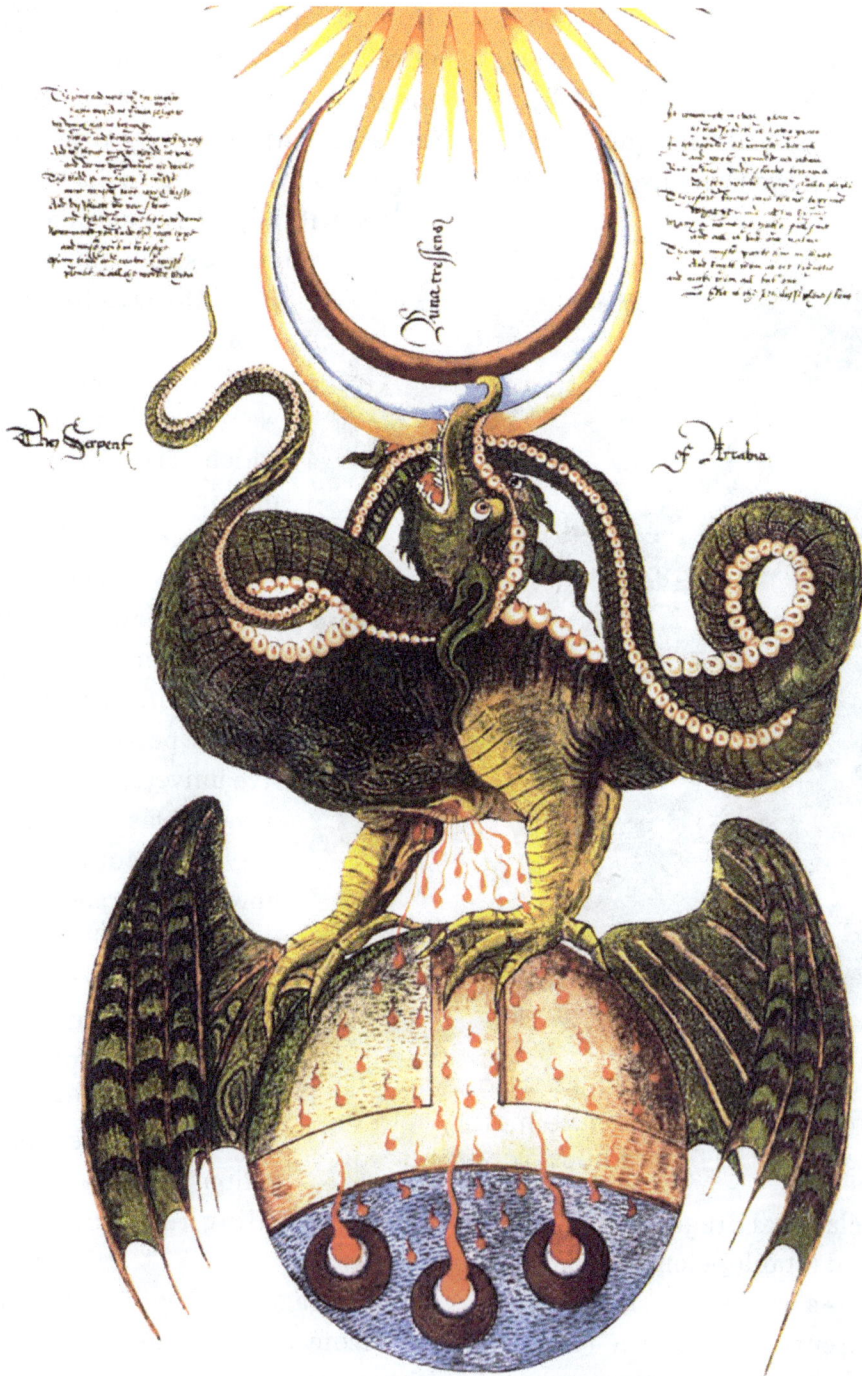

A spectacular lindorm from the 15th-Century
alchemical scrolls of Sir George Ripley.

devouring and destroying each another symbolised purification, and if they then gave rise to a so-called 'glorified dragon', this was the longed-for Lapis Philosophorum—the Philosopher's Stone. A group of dragons fighting each other personified the gradual separation and transmuting of elements via putrefaction (nigredo); or, on a spiritual level, psychic disintegration.

An alchemical wyvern ouroboros, from Michael Maier's book *Atalanta Fugiens*, 1617.

There were even some dragons so unusual, such as two-limbed amphisbaenas with wings, that they rarely if ever featured in myths and legends but appeared as alchemical symbols because their precise forms personified highly-specific processes or spiritual states. And the bizarre pseudo-religious image of a dragon nailed to a Cross actually signified a specific alchemical process involving the fixation of the volatile element. In short, the dragons of alchemy were analogous to the physical reactions of modern-day chemistry as well as to psychological states and activities.

Alchemical recipes often included highly exotic-sounding ingredients, one of which was dragon's blood, said to be extremely toxic. Indeed, the blood of the vishap, an Armenian water-associated classical dragon living on Mount Ararat, was so poisonous that any weapon dipped in it would kill instantly.

An engraving from 1854 of the great dragon tree of Orotava, Tenerife.

In reality, this rare, highly-prized substance was the bright-red resin from a very large, peculiar tree-like plant called the dragon tree *Dracaena draco*, native to the Canary Islands, Madeira, Cape Verde, and Morocco.

DRAGONS IN HERALDRY

The dragon is a very popular symbol in heraldry too, but just as in alchemy it appears in many different forms.

The winged classical dragon is generally depicted nowadays in heraldry with a barbed tongue and tail (though in earlier times the tail was generally smooth), large fully-veined bat wings, upperparts that are sometimes scaled, roll-armoured underparts, ears of varying shapes and sizes, a sharp tusk protruding upwards from its nose, and clawed feet. It symbolises power, strength, invincibility (especially in battle), and—unusually for Western dragons—good fortune.

The precise position (attitude) in which this dragon (and other quadrupedal beasts) is depicted has a range of specific heraldic terms associated with it. Represented looking left in profile while standing on all four feet, its pose is described as statant. If looking left in profile while standing with one of its forelimbs raised, this pose is passant. If in the same pose as passant but with head turned to look back upon itself, this is passant reguardant. If looking left in profile while standing upright on one hind leg (more rarely both) but with both forelegs raised, this is rampant. If in the same pose as rampant but looking at the observer (instead of in profile), this is rampant guardant; if looking left in profile while sitting upright on its haunches, this is sejant. If in the same pose as sejant but with its forelimbs raised, this is sejant erect. If looking left in profile while lying down with head raised, this is couchant, or dormant if the head is down with eyes closed.

Known formally as Y Ddraig Goch or the red dragon of Cadwallader (a 7th-Century Welsh king who adopted it for his standard), the red dragon with yellow underparts that appears as a supporter in the Coat of Arms of England, and also in those of the English Tudor kings Henry VII, Henry VIII, and Edward VI, as well as (in entirely red form) upon the Welsh flag, is the most famous heraldic winged classical dragon. In rampant attitude, it also features in the coat of arms for the French city of Plomelin in Finistère, Brittany. However, it is just one of many variations upon the theme of this dragon type.

A heraldic winged classical
dragon in passant pose.

When Elizabeth I ascended the English throne in 1558, for instance, she changed the colour (tincture) of her coat of arms' winged classical dragon supporter from red (known as gules in heraldry) to gold (known as or); but when the Stuart dynasty ascended the English throne upon her death in 1603, the dragon was replaced by the Scottish unicorn. The coat of arms of Bad Goisern in Upper Austria features a black (sable) dragon with red jaws, tongue (forked), and claws, looking back upon itself (reguardant). The coat of arms of Frasdorf, a municipality in the Upper Bavarian district of Rosenheim in Germany, features a silvery white (argent) dragon looking right in couchant attitude.

A number of noble families in Russia and Poland have coats of arms portraying St. George mounted on horseback spearing a dragon, often green (vert) in tincture. This image also appears in the Regalia of England's Order of the Garter, which was founded in 1348 on St. George's Day by Edward III, and is the earliest Order of Chivalry on record from anywhere in Europe.

In German heraldry, a wingless quadrupedal classical dragon is known as a lindwurm, whereas, confusingly, in British heraldry the term 'lindworm' is applied to a wingless two-legged dragon (which, technically, is a lindorm). Winged two-limbed dragons, i.e. wyverns, are very popular heraldic symbols, representing power, strength, and endurance, and many German towns include either a two-limbed wingless dragon or a wyvern of various colours in their coat of arms.

Indeed, the wyvern is the most commonly-represented type of dragon in heraldry, but long before this it was already a symbol closely associated with the ancient

Saxon kings of Wessex. A golden wyvern featured on the flag of England's last Saxon king, Harold II, whose army was defeated at the Battle of Hastings by William the Conqueror's Norman invasion, and it also appears in the Bayeux Tapestry recording this highly significant event.

A heraldic wyvern. (Huber, 1981)

One of the most famous pairs of wyverns in British heraldry appear as supporters in the Duke of Marlborough's coat of arms, in which the wyverns are each depicted sitting erect upon its coiled tail with its claws in the air. Two wyverns also serve as supporters of the shield of the Dukes of Rutland. Perhaps the most unusual heraldic wyvern, however, features in the crest of Maule—green in tincture, it has two heads, vomiting fire at both ends.

A heraldic cockatrice. (Huber, 1981)

Less popular but still represented in some coats of arms is the cockatrice, which is depicted as a wyvern but with the head of a cockerel (rooster) rather than a dragon, and whose beak, coxcomb, and wattles are often in a different tincture from the rest of it. Two cockatrices act as supporters on the Earl of Westmeath's coat of arms.

A seven-headed winged hydra appears in the crest of various French families, including Barret, Crespine, and Lownes. The salamander is best known as the personal device of Francis I (1494-1547), king of France, giving rise to the city of Paris's coat of arms, but it has also featured in the ironmongers' coat of arms since 1455.

Other dragon types occurring in heraldry include the amphiptere (although generally confined to crests rather than shields), the basilisk, and the sea-dragon. Even the Chinese dragon occasionally appears, as the dexter (right-hand) supporter in Baronet Hart's coat of arms, for instance, and in the Berkshire Regiment's badge.

DRAGON TATTOOS

Body art has experienced an enormous resurgence in recent years, especially tattoos and tattooing, and one of the most fashionable, in-demand subjects for a tattoo is the dragon. Indeed, there is even a much-sought-after book devoted entirely to dragon tattoos—Donald Ed Hardy's lavishly-illustrated tome *Dragon Tattoo Design* (1988), containing countless designs for every taste in dragons.

Whether in the West or the East, most tattooed dragons are of the Oriental category (though the red Welsh dragon is popular as a symbol of Celtic patriotism). Its slender, serpentine form and rich spectrum of vivid colours allows itself to be readily applied to a person's back, arms, chest, or legs, and to achieve individuality for them. Several notable celebrities sport dragon tattoos, including singers Lenny Kravitz, Pink, and Mels B and C of the Spice Girls, as well as actor Bruce Willis, and actress Angelina Jolie.

There is a plethora of symbolism associated with dragon tattoos, depending in particular upon their type, colour, activity, the presence of other animals alongside them, and the sex of the person selecting them.

A Western dragon in a tribal tattoo.

Oriental dragons in general are associated with freedom, protection, and nobility, but in keeping with their status and role in traditional mythology, the tattoo of a Chinese dragon more specifically symbolises good fortune and wisdom, whereas the more elongate, fewer-clawed Japanese dragon personifies balance in life. As already noted, the Celtic dragon symbolises pride in one's Celtic ancestry but also power and strength, and is sometimes depicted alongside a crown or even a throne. Other Western dragons, which are particularly popular when portrayed as stylised tribal tattoos, also represent bravery, ferocity, and even war-like attributes—hearkening back to their traditionally darker, less benign image than that of their Oriental counterparts.

A tattoo of a yellow Oriental dragon symbolises knowledge, helpfulness, and a good companion, as does one of a gold Oriental dragon (which is also utilised to indicate a kind personality), whereas a green one embodies life and the planet, a red one symbolises keen eyesight, and a black one shows that its bearer has old but wise parents. A blue dragon tattoo is associated with compassion and forgiveness. But beware: in Chinese tradition, blue dragons also symbolise sloth and idleness—so make sure that you don't give out the wrong impression if you choose this symbol as a tattoo!

A tattoo depicting an Oriental dragon protecting treasure signifies wealth (either material or spiritual). A coiled dragon personifies the oceans but also suggests that, just like those vast bodies of water, the person bearing this tattoo has hidden, mysterious depths to their personality, and a horned dragon represents strength and authority in both deed and intent, because only the more advanced Oriental dragons possess horns. If the fuku-riu type of Japanese dragon is chosen for a tattoo, its bearer is hoping to attract good fortune, as this type is traditionally the luck dragon in Japan.

Oriental dragons are still revered as rain and water deities, so if such a dragon is tattooed reposing near to water, i.e. its normal resting state, this symbolises tranquillity and peace of mind. However, if the dragon's teeth show, or, if winged, its wings are extended, this can denote hostility or aggression. Taking great care and giving thought beforehand when selecting a dragon tattoo is very important, therefore, to avoid its sending out an inappropriate or unwanted message, especially as tattoos cannot be easily removed or amended once applied.

Dragon tattoos on men typify the general dragon-allied symbolism of power, wisdom, courage, and protection, and are often applied to readily-visible regions of the body like arms or legs (as sleeve tattoos), across the shoulders or chest, or encompassing the back. When present on women, conversely, they tend to appear on less overt regions, such as feet, ankles, nape, near the navel, or down the sides, and are more closely linked with creation, emphasising that it is women who give birth. Consequently, mothers (especially new ones) often select a dragon when choosing a tattoo.

Sometimes an Oriental dragon and Chinese phoenix are tattooed together, forming a circle or otherwise intimately linked with one another, which symbolises a successful marriage. Tigers in ancient Oriental tradition often represent aggression and evil intent, so a dragon tattooed above a tiger indicates that its bearer has overcome darkness in their life, or intends always to do so. Avoid tattoos of dragons battling with tigers if tattoo symbolism is important to you, however, because these images can indicate aggression or internal conflict.

Tattoo of a serpent dragon entwined around a dagger, symbolising strength. (Dr. Karl Shuker)

An attractive way of enhancing both the physical appearance and the symbolic significance of a dragon tattoo is to add a message alongside it, written in either Chinese Hanji or Japanese Kanji script.

Above all, a dragon tattoo signifies that the bearer is special. If there could be just a single tattoo to personify the phrase "Why be ordinary when you can be extraordinary?", it would be a dragon.

DRAGONS AND FENG SHUI

According to time-honoured Chinese beliefs, ancient megaliths such as stone circles, standing stones, and burial mounds, as well as sacred temples and other once- or still-venerated monuments, are linked to one another not just in China but throughout the world via lines of earth energy known as lung mei or dragon paths (in Britain they are called ley lines, in Australia song lines, etc). And the reason why these edifices are positioned precisely where they are is believed to be that they are acting as barriers to this energy, diffusing it, because otherwise it would be too powerful and could therefore be dangerous to life. Harnessing and controlling the flow of earth energy is known in China as feng shui, and the good, life-enhancing form of earth energy flow is called qi or sheng chi, which is also the essential life-force of all living things.

This scenario is not exclusive, however, to large-scale, outdoor topography. It also occurs in relation to the smaller yet no less significant energy flow that occurs around and inside a person's home. But rather than utilising full-sized religious monuments to block or channel the dragon paths, and attract only the positive sheng chi energy into the home in order to induce good fortune, smaller yet no less potent figures are used instead. The most common of these are ornaments or figurines in the shape of an Oriental dragon, usually holding a pearl with one of its clawed feet, which symbolises wealth, power, and plentiful opportunities.

The colour, location, and number of dragon figures inside a home all exert significant effects upon its internal energy flow. For instance, the presence of a golden dragon is auspicious for attracting wealth, whereas a green dragon promotes health—but all of this in turn depends upon the dragon's specific location.

One of the author's feng shui dragons. (Dr. Karl Shuker)

According to the tenets of feng shui, a dragon figure should be placed at or below eye level (to keep it under control), in or near to an open space (to maintain its power), not facing a closed wall or door (which would restrict its power), and not with its pearl-holding claw facing an open door or window (this would encourage your good fortune or wealth to escape through that opening). Dragons should not be placed in areas of the home containing low levels of energy, such as closets, the bath-room, or garage, as this will waste or pollute their powers of attracting positive energy. And it is very important not to diminish a dragon's power in any way; the maximum number of dragon figurines recommended for a home is no more than five.

The author's vintage feng shui bixi. (Dr. Karl Shuker)

Placed in the home's eastern corner, a dragon stimulates longevity and well-being within the family, and good fortune in the home owner's personal career will occur if one is placed in a northern location or in the owner's personal corner of their living room, dining room, or office. If placed close to an aquarium, fountain, or other water source, a dragon will attract good fortune for enhancing the home owner's career or chances of fame. Dragons made of metal are said to be effective in fending off bad energy responsible for illness, accidents, and violent confrontations.

A second type of dragon figurine popularly utilised in feng shui within the home

The author's feng shui Chinese phoenix from Hong Kong. (Dr. Karl Shuker)

is the bixi or tortoise dragon—a curious composite beast combining the body of a tortoise with the head of an Oriental dragon. Of particular benefit to academic persons living or working in the home, the tortoise dragon should be placed in the home's northeast corner or upon an office desk or work table in order to promote success in examinations or other educational tasks. This will be especially beneficial if the home owner was born in the Year of the Dragon—the dragon being the fifth sign of the Chinese zodiac. If a figurine of one of these creatures is placed facing the home's front door, this should encourage an untroubled existence inside the home by warding off danger or troubles that might otherwise seek entry.

Two other symbolic creatures of great significance in feng shui are the tiger and the Chinese phoenix. Sometimes these are aligned with one or other of the dragon symbols to achieve a particular outcome.

One of the most widely used examples is pairing a dragon figurine with a phoenix figurine in a home's bedroom, because the masculine (yang) power of the dragon will be matched by the phoenix's feminine (yin) power, and together constitute a potent symbol of marital contentment. If, conversely, they are not paired and the bedroom is shared by a man and a woman, the single unpaired dragon's power or the single unpaired phoenix's power will be nullified respectively by the yin emanating from the woman or the yang from the man.

CHAPTER 6:
DRAGONS IN THE VISUAL ARTS

THROUGHOUT HISTORY, dragons have been closely linked with the visual arts, but there are certain periods and examples that are particularly associated with these iconic reptiles, as readily demonstrated by the varied selection examined here, from the past and the present.

DRAGONS IN ANCIENT ART
As far as human civilisation is concerned, there has never been a time without dragons, because artistic representations of these fearsome if fictitious saurians date back as far as early humanity itself. Indeed, the oldest known depictions of dragons, on pottery, were created as long ago as 6200-5400 BC, during the Neolithic Period, by northeastern China's Xinglongwa culture.

Dragons and dragon hybrids were popular subjects for portrayal on drinking vessels by the Zhaobaogou culture (5400-4500 BC), which succeeded the Xinglongwa. And the Hongshan culture (4700-2900 BC) produced beautiful dragon sculptures from jade (constituting some of the earliest known examples of jade working). These included elegant figurines of dragon embryos, and

pig dragons (with boar-like snouts and coiled, limbless, elongate bodies), some of which have been discovered sealed inside tombs.

In 1993, archaeologists announced the discovery of the oldest known painting of a Chinese dragon. This was a cave painting found on a cliff in the southwestern corner of Shanxi province in north-central China, surrounded by images of deer and other animals, and also associated with over a thousand stone tools, all of which dated back approximately 10,000 years to the Mesolithic Period.

In the Middle East, Apep was a giant, sometimes-winged serpent dragon of the sky that attempted each night to devour the sun in ancient Egyptian mythology, and is known from depictions dating back as far as 4000 BC. Many famous subsequent paintings of it are also known, such as the version on a wall of the tomb of Inherkha at Deir el Medinah, dating from c.1164-1157 BC. Here Apep is slain by the sun god Ra in the guise of a huge cat.

Ancient Mesopotamia was another plentiful source of early dragon art in the

Middle East. Most famous is the spectacular Ishtar Gate, created in c.575 BC by order of Babylon's King Nebuchadnezzar II, and depicting the three royal creatures—the

Tomb wall depiction of Apep being slain by Ra in the guise of a cat.

bull, the lion, and the dragon or mush-ussu—set in rows and portrayed in gold amid the gate's brilliant cobalt-blue brick-work. Even older, however, dating from c.800-600 BC, is the impressive bronze sculpture of a horned dragon's head that was excavated in Mesopotamia and is now housed at the Louvre in Paris. Older still is a relief dating from the ruling Kassite dynasty (c.1530-1150 BC) of Babylonia that depicts the dragon-slaying sun god Marduk

standing alongside a small scaly dragon lying with horned head raised. Other notable dragon artefacts from ancient Mesopotamia include Babylonian cylinder seals portraying the primal deity Tiamat in her mighty sea dragon guise.

RENAISSANCE DRAGONS

The European Renaissance Period of cultural reawakening spanned the 14th to the 17th Centuries, and produced some of the most sumptuous and spectacular works of art ever created, including a rich spectrum of paintings, drawings, engravings, and sculptures, as well as tapestries, stained-glass windows, murals, altar works, illuminated manuscripts, and illustrations for bestiaries. Popular subjects depicted included classical legends as well as Christian subjects, and dragons were well-represented.

Perhaps the single most popular dragon-related theme portrayed by Renaissance artists was the future St. George battling the dragon. Of particular interest is the great variety of dragons so depicted—winged and wingless classical dragons, serpent dragons, lindorms, wyverns, dragonets, and even multi-headed varieties have all been pictured as the saint's reptilian aggressor.

Two of the most famous paintings of this legendary confrontation are by Italian artist Paolo Uccello (1397-1475), which

One of Paolo Uccello's famous paintings of St. George slaying the dragon.

portray two different scenes but both featuring the dragon as an ocellus-winged wyvern, with wings spread, displaying their conspicuous eye-like markings. These are often seen on the wings of butterflies and other insects, serving to distract or shock a would-be predator, and were possibly added by Uccello to indicate a similar tactic utilised by the dragon in an attempt to ward off the saint. Other well-known wyvernesque versions include those portrayed by Flemish painter Rogier van der Weyden (1400-1464) and Venetian painter Paris Bordon (1500-1571).

The four-legged winged specimen attacking St. George and his steed as painted by Vittore Carpaccio (1465-1525/6) in his masterpiece from c.1502 is oddly canine in general form, especially its head, whose jaws are being skewered by the saint's lance. Even more so, however, is the version by German painter Jost Haller (c.1410-c.1485), in which the quadrupedal dragon is not only wingless and scaleless, but so disconcertingly mammalian in overall appearance that St. George's spearing it through the neck and its agonised expression are quite disturbing to view.

Religious stained-glass windows commonly portray St. George's antagonist as a serpent dragon. This is probably because it is easier to depict in glass a limbless creature rather than a more complex form with limbs, talons, tail, and wings.

One aspect little commented upon but curious nonetheless in relation to such

Vittore Carpaccio's magnificent painting of St. George and the dragon.

Stained-glass window at the author's home, portraying St. George and his defeated scaly foe, which is depicted here as a serpent dragon. (Dr. Karl Shuker)

works of art is the remarkably small size of the dragon in many of them, sometimes no larger than a dog and therefore more correctly categorised as a dragonet rather than as a fully-fledged dragon. Perhaps this is compensated for by the fact that in European myths, dragonets were often disproportionately dangerous, especially the venomous ones. Alternatively, the artists may have been expressing visually that good is always greater than evil, or they could have simply been giving St. George a not so subtle advantage! Thankfully, however, there have been some suitably sizeable dragons too, exemplified by the huge saurian foe bravely dispatched by St. George in the vivid portrayal by Lucas Cranach the Elder (1472-1553).

St. Michael slaying the dragon (representing the devil) was another much-depicted Christian theme during the Renaissance, with a similar degree of morphological diversity exhibited by the dragon. In keeping with the scriptures' description of the Dragon of the Apocalypse, however, a multi-headed version was a popular choice—as epitomised by the exquisitely detailed paintings and engravings of German artist Albrecht Dürer (1471-1528).

One notable, pre-Renaissance exception to this trend, however, is a famous, colourful illustration in the *Liber Floridus* (an early encyclopaedia compiled between 1090 AD and 1120 AD by Lambert, Canon of Saint-Omer), in which the dragon battling St. Michael and the angels is a very large wyvern, standing erect with bat-wings spread. Another exception, albeit from the pre-Raphaelite period, is a stirring portrayal of this biblical battle by Sienese painter Ricciardo Meacci from 1897, in which the dragon is single-headed and furry-chested. One of the strangest representations, however, appears in a depiction

St. Michael and the angels battling the Dragon of the Apocalypse portrayed as a wyvern in the *Liber Floridus*.

housed in Barcelona's Museum of Catalan Art, showing St. Michael battling a wyvern whose tail terminates in four separate dragon heads.

Among the most portrayed of Greek myths featuring a dragon is Perseus rescuing the princess Andromeda from the great sea dragon Cetus, whose appearance has been the subject of some very remarkable interpretations. A glorious painting depicting this dramatic scene is that of Piero di Cosimo (1462-1521), produced in c.1510-13 and entitled *The Liberation of Andromeda*, in which the dragon is a surprisingly walrus-like entity, albeit also winged and flailing a long corkscrewed tail.

Even more incongruous is the hound-headed, modest-sized sea dragon in a version of this same scene painted in 1602 by Giuseppe Cesari (1568-1640). Much later, but equally odd, is the winged classical dragon representing Cetus in Frederic Leighton's painting from 1891, despite Cetus supposedly being a marine dragon sent by Poseidon. In a painting by French neoclassical artist Jean-Auguste-Dominique Ingres (1780-1867), conversely, it is depicted as a bona fide maritime serpent dragon.

Cadmus slaying the giant serpent dragon guarding the Castalian Spring at Boeotia is another much-illustrated Greek myth. However, it probably attained its visual zenith via the stunning pen and ink drawing executed in 1588 by Dutch artist Hendrik Goltzius (1558-1617), entitled *The Dragon Devouring the Fellows of Cadmus*.

Other popular subjects for Renaissance artwork are the twelve labours of Heracles,

Piero di Cosimo's walrus-like sea dragon Cetus
in his painting *The Liberation of Andromeda*.

one being the slaying of the Lernean hydra. A decidedly scrawny specimen was clubbed senseless by the hero in the painting by Italian artist and sculptor Antonio del Pollaiolo (1432-1498). Fortunately, more formidable depictions of this polycephalic lindorm also exist, such as the robust portrayal by Spanish painter Francisco de Zurbarán (1598-1664), as well as various post-Renaissance examples, like the vibrant engraving by Bernard Picart (1673-1733), and a truly exquisite portrayal by American artist John Singer Sargent (1856-1925).

Not even geniuses of the calibre of Leonardo da Vinci (1452-1519) or Michelangelo (1475-1564) were immune to the magical allure of dragons. One of the former artist's most impressive if lesser-known works is a very powerful drawing of a huge wyvern with wings apart attacking a lion; and in c.1525 Michelangelo produced a detailed chalk and paper drawing of a wonderful coil-necked, plume-winged dragon.

THE DRACONOPIDES—A DRAGON IN THE GARDEN OF EDEN?

One of the most extraordinary dragons has never appeared even in myths, legends, and folklore, let alone in the real world. Instead, and very remarkably, this curious being, known variously as a draconopides, draconcopedes, or draconipes, was cultivated primarily within the art world, as a device for representing one of the most controversial entities in the Christian world—the serpent

Produced in 1921, John Singer Sargent's sumptuous rendition of Heracles battling a mighty 13-headed hydra.

that tempted Adam and Eve in the Garden of Eden.

According to the Bible, after betraying Adam and Eve the serpent was cursed by God to crawl limbless on its belly in the dust thereafter. As recognised for centuries by theologians, however, this begs the obvious question: what did the serpent look like *before* it was cursed?

Although early painters depicted it as a normal snake coiled round the Tree of Knowledge, scholars recognised that representing it like this posed problems, because it had clearly been something much more noble originally. Indeed, some early texts claimed that it had been closer in form to humans than to its fellow creatures. By the 12th Century, European artists had begun

to adopt a very different but relatively standardised appearance for the pre-cursed serpent—which became known as the draconopides.

Whereas its body was entirely serpentine, lacking limbs, and normally depicted wrapped around the trunk or branch of a tree, the head of this most intriguing entity was human, complete with hair and an expressive female face. In subsequent times, moreover, the draconopides evolved into an even more elaborate, composite being.

Its head and face were still those of a woman, and although it retained its serpentine body, the upper region of it had now differentiated into that of a woman too, complete with breasts, and it also possessed a pair of human arms and hands. Sometimes it even sported a pair of bat-like wings too, but did not possess legs or feet, in spite of its name—ironically, 'draconopides' translates as 'dragon-footed'!

Dating from c.1473, one of the most famous depictions of this new-improved Renaissance draconopides appears in François Fouquet's painting *Le Péché Originel* (*The Original Sin*—a title also used by several other artists for their versions of this same scene). And a similar but wingless counterpart appears in the temptation scene portrayed in one of the panels constituting Michelangelo's glorious series of paintings on the ceiling of the Vatican's Sistine Chapel (1508-12).

Other portrayals of draconopides were produced by the likes of Benjamin the Scribe (fl. 1280s), Masolino da Panicale (c.1383-c.1447), Raphael (1483-1520), and Cornelis van Haarlem (1562-1638). And a spectacular relief featuring Adam, Eve, and a draconopides had been sculpted during 1425-38 by Jacopo della Quercia (1374-1438) for Bologna's San Petronio church. Draconopides have also appeared in some stunningly beautiful stained-glass windows, such as the exquisite example in Germany's Ulm Cathedral that dates from 1420.

A draconopides became a favoured guise among artists for portraying Lilith too. According to Rabbinical lore, she was Adam's first wife, before God created Eve, but deserted him after refusing to be subjugated by him, becoming a demon instead. Moreover, in some texts she was even synonymised with the Eden serpent, so that it was Lilith who tempted her former husband and his new wife.

In modern times, however, the implications of how God's curse had changed the serpent's appearance from its original form appear to have been lost upon the art world. For contemporary renditions of this creature in the Garden of Eden have reverted to portraying it simply as a typical limbless snake. RIP the draconopides.

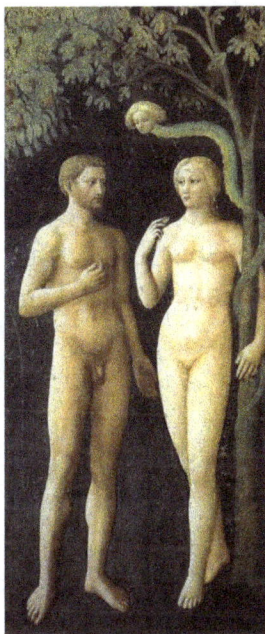

The Temptation of Adam and Eve by Masolino da Panicale, c.1425, a fresco in Brancacci Chapel at the Church of Santa Maria del Carmine, in Florence, Italy.

DRAGONS ON MAPS AND GLOBES

The Latin phrase "Hic sunt dracones" ('Here be dragons') is widely attributed to early cartographers, utilising it on maps and globes in order to denote any unexplored or inhospitable expanse of land. In reality, however, only one confirmed example of such usage is known—on the Hunt-Lenox Globe (c.1503-1507), a hollow copper globe of the world. Purchased in Paris by architect Richard Morris Hunt in 1855, then given to American bibliophile James Lenox, it is of unknown origin, but now resides in the New York Public Library.

In contrast, visual depictions of dragons and marine dragons or sea serpents appear on several important cartographical representations. Perhaps the best known is the 'Carta Marina' map of Scandinavia (1539), prepared by Swedish scholar Bishop Olaus Magnus (1490-1557), which portrays a vivid array of sea monsters as well as a wyvern-like or cockatrice-like beast in northern Lapland.

DRAGONS IN FANTASY ART

By the end of the Renaissance Period, the dragon's popularity in art was in decline, and it remained out of the limelight until the 20th Century. True, there were some magnificent exceptions, such as the extremely potent, vigorous painting produced in 1788 by Swiss-originating British painter Henry Fuseli (1741-1825) of Thor battling the Midgard serpent; the heroic slaying of Fafnir by Siegfried vividly portrayed in 1880 by Konrad Dielitz (1845-1933); and the celebrated watercolour and engraving by

William Blake's famous illustration of Behemoth (top) and Leviathan (bottom).

mystic painter-poet William Blake (1757-1827) of the Bible's two colossal monsters, Behemoth and Leviathan, for his *Book of Job* (1826). In general, however, the dragon's day for inspiring art seemed largely over.

During the late 1800s, however, due in no small way to major advances in offset litho-

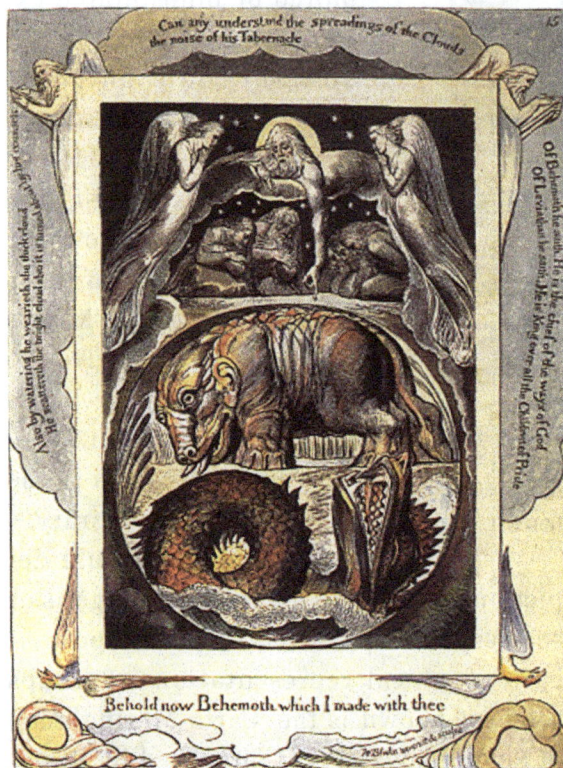

A classic Walter Crane illustration for the French fairy tale 'Princess Belle-Etoile,' depicting Prince Cheri battling a three-headed dragon.

graphy and other publishing techniques, a second renaissance of dragon art occurred, which has flourished ever since. It began when children's illustrated books came into favour, many of which contained retellings of dragon-related myths and fairy tales, and thus needed pictures to accompany them.

One of the premier proponents in this specialised field was the highly influential English artist and book illustrator Walter Crane (1845-1915). Another prolific illustrator of such books during this same period was Arthur Rackham (1867-1939)—within whose vast output of artwork for children's literature were several dragons, most notably his ferocious multi-fanged

Leviathan for a book of fairy stories. So too was Henry Justice Ford (1860-1941), whose stunning images for Andrew Lang's beloved series of twelve *Rainbow Fairy* books (1889-1910) included some superb dragons and monstrous serpents.

Arthur Rackham's Leviathan.

In addition, the new literary genres of science fiction and 'swords and sorcery' fantasy, arising at much the same time and swiftly burgeoning during the early 1900s via countless novels and magazine articles, required eyecatching illustrations for their covers, dustjackets, and plates (if illustrated internally too). Suddenly, depicting dragons was big business, and an entire new industry was launched—fantasy art.

Today, there are quite literally hundreds of famous fantasy artists worldwide, many of whom have incorporated dramatic portrayals of dragons within their repertoire, including the following necessarily brief selection of noteworthy examples.

And where better to begin than with John Howe? Born in Vancouver, Howe's most significant of numerous highly-merited claims to artistic fame is that it was his portrayals of them as dragonesque creatures that shaped movie director Peter Jackson's vision of the fell beasts, the foul winged steeds of the nine Nazgûl or Ringwraiths, in his epic *Lord of the Rings* film trilogy.

Famous for a visual phantasmagoria of dragons down through the years, everything from winged fire-belchers to multi-finned sea-colossi, the Brothers Hildebrandt—Greg and Tim—were twin brothers from Detroit, Michigan, who had collaborated on numerous projects since 1959, such as calendars, book covers, posters, comics, trade cards, and much else besides. They became famous for the ultra-realism of their work, but, tragically, Tim died of diabetic complications in 2006, aged 67.

Another name intimately associated with fantasy art is Boris Vallejo. Born in Peru in 1941, he emigrated to the USA in 1964, and has established himself as a much sought-after illustrator of covers for sci-fi and fantasy novels. Often quite erotic in appearance and sometimes modelled upon himself and his wife Julie Bell (another acclaimed fantasy artist), his artwork characteristically features muscular male

A delightfully droll piece of dragon-inspired comical artwork by Swedish fantasy artist/film-maker Richard Svensson, entitled *Silly Fantasy*. (Richard Svensson)

barbarians and Amazon-type female warriors battling a vast assortment of ferocious monsters, including various types of dragon. Perhaps his most idiosyncratic yet famous dragon-featuring painting, however, is 'Tattoo,' which depicts in incredibly life-like form a man's tattoo of an Oriental dragon coming to life upon his heavily-muscled forearm and biting it, drawing forth a crimson trickle of blood.

With the advent of digital art, some of today's most accomplished fantasy artists use computers instead of paint brushes to create their illustrations. Among the most celebrated exponents of this modern-day technique are the likes of Photoshop-utilising graphic art experts Andy Jones, Kerem Beyit, and Scott Altmann. One of the most stunning dragon paintings created digitally is Altmann's *Rodin* (2008), in which an immense swamp-dwelling serpent dragon rises from the water to view the cautious approach of a small gondola-like boat bearing a single hooded human figure with arms open in supplication.

Ciruelo Cabral, an Argentinian painter who specialises in depicting dragons, is one of the most original artists in the fantasy genre, not just on account of his artwork but also due to what he paints it upon. This is because his preferred medium is neither canvas nor computer screen, but consists instead of stones and rocks that he has picked up, and whose specific shapes, forms, and textures directly influence his choice of dragon images to be painted upon them.

DRAGON-INSPIRED MEGA-STATUES

Architectural dragons can be found in profuse quantities in the West—intricately carved within elaborate panels and misericords inside churches and cathedrals or as gothic water-spouting gargoyles outside them; as imposing statues in parks, squares, and thoroughfares in every major Western capital city; and even as weather vanes, ornamental fountains, and bridges. They are equally numerous in the East too—snarling down in gilded splendour from roof beams and doorways, coiled and spiralling around columns and pillars, and emblazoned in glowing colours upon richly-ornamented screen walls.

But for truly enormous statues of dragons, we need to look elsewhere—to theme parks. At least three different parks claim records concerning the size of certain dragon statues there. Opened in 2008, Siam World on Tenerife in the Canary Islands is said to be the world's largest water park, and houses a gargantuan statue of a Thai dragon that the park believes to be the world's biggest. In the Suoi Tien Amusement Park, a huge Buddhist theme park in Vietnam's Ho Chi Minh City (formerly Saigon), there is a 1312-ft-long dragon statue said to be the largest anywhere in southeast Asia.

According to the record books, the world's tallest dragon statue can be found inside the Xieng Khuan (aka Buddha Park), a sculpture park located 15 miles southeast of Vientiane, the capital of Laos in southeast Asia, and publicly opened in the mid-1970s. Standing 82 ft high and composed of reinforced concrete, this colossal work of art depicts the Lord Buddha meditating with an enormous seven-headed naga upraised behind him in protective mode.

How much longer the above mega-statues will lay claim to their titles, however, remains to be seen, because they face serious competition from two very recent challengers—if they are ever completed, that is. They hail respectively from China and Wales—two countries renowned for their longstanding dragon links.

Officially revealed to the world's media in 2007, when its construction was already well underway (having begun in 2002), China's giant dragon statue—formally known as the *Ancestral Dragon*—would unequivocally be the world's biggest ever, a truly stupendous creation in its final form. Situated close to Xinzheng (part of the major city of Zhengzhou) in Henan Province, composed of concrete and marble, and with an estimated total cost of $300 million, it would measure 13 miles in length when complete, snaking along the ridge line of Shizu Mountain, north of the city. In official Chinese press-release photos released in 2007, its partially-completed horned head, enshrouded by scaffolding, rose 30 ft above the ground, but its final height would be almost 100 ft. Moreover, the finished dragon would be covered in 5.6 million scales of jade or gold-coated bronze.

There were also ambitious plans to build a vast complex of galleries, shops, restaurants, elite clubs, and other tourist attractions inside the dragon's hollow body (30 ft tall, 20 ft wide), as well as installing a light rail system there for transportation.

Buddha Park's giant naga statue. (Jpatokal/Wikipedia)

Its instigators and designers hoped that everything would be complete by 1 October 2009, in time to mark the 60th anniversary of the People's Republic of China's founding.

Sadly, however, the much-needed funding from Chinese businesses and other potential sources of private investment never materialised, and the vast majority of people surveyed for their views concerning the statue announced that they didn't want it, denouncing it as a waste of money and detrimental to the environment. And so, over five years later, China's *Ancestral Dragon* remains unfinished, lying abandoned and uncherished—a far cry indeed from the veneration normally accorded to dragons in the East.

In the West, meanwhile, an equally ambitious project featuring a dragon statue was launched in February 2011. This was when the Welsh Dragon Project secured official planning permission to erect in Wrexham, North Wales, a 134-ft-tall steel and glass visitors' viewing tower, on top of which would sit a 75-ft-tall Welsh dragon cast from bronze and boasting a wingspan of more than 150 ft (bigger than that of a Boeing 737 aeroplane!).

Entitled *Waking the Dragon*, this spectacular monument would be the tallest public artwork in the United Kingdom, even overshadowing Nelson's Column in Trafalgar Square, London. Alongside it would be a centre devoted to the history and culture of Wales, plus a museum, bar, and restaurant. The brainchild of art dealer Simon Wingett, *Waking the Dragon* will cost an estimated £6 million, but Wingett hopes to obtain the entire cost exclusively from commercial sponsorship. Plans for this impressive project to be complete in time for the Summer 2012 Olympics held in London were not met, however, and it remains uncertain when—or if—this monumental Welsh dragon will ever raise its wings above Wrexham.

Dubbed 'the world's first theme park,' the Garden of Bomarzo, located in Viterbo, northern Lazio, in Italy, is more commonly known as the Park of the Monsters—and for good reason. Created by Duke Pier Francesco 'Vicino' Orsini (c.1523-c.1583), an ex-military officer who was also a leading patron of the arts, as a tribute to his late wife, Giulia Farnese, this Renaissance wonderland is populated by over two dozen immense and often very macabre statues. These include a three-headed hell hound, a giant ripping apart another giant, a Carthaginian war elephant carrying a trampled Roman soldier aloft in its trunk, and, most spectacular of all, a huge dragon holding at bay a dog (symbolising spring),

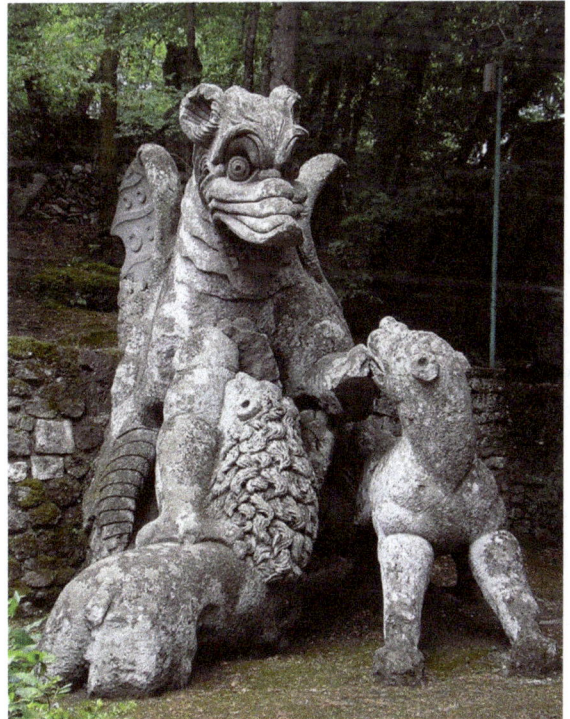

Dragon sculpture at Bomarzo's Park of the Monsters. (Silvana Pellegrini Adam)

a lion (summer), and a wolf (winter). Following the duke's death, the park fell into disrepair and became wholly overgrown, but in 1970 a successful restoration was initiated by the Bettini family owning the land containing this most surreal of gardens, and today it is a very popular if exceedingly strange and eerie tourist attraction.

DRAGONS IN THE MOVIES

With the coming of the movies, the visual arts—and dragons—came alive! For the world of the cinema is also one of dragons, threatening to burst forth from the big screen into the real world at any moment, especially with new advances in 3-D cinematography! Thankfully, however, they never quite succeed.

Instead, we can experience from the safety of our own armchair or the cinema seat the vicarious thrill of modern-day knights confronting their deadly reptilian foes, masterfully engendered by the imagination and artistry of animators or the technical wizardry of live-action and CGI adepts. Their skills enable us to enter realms of fantasy and virtual reality that at least for the running time of the film are no less vital than our own world—and sometimes are even more so. Here, anything is possible—even dragons.

DISNEY DRAGONS AND OTHER BIG-SCREEN CARTOON VERSIONS

Given the limitless possibilities of expression available in animation, the dragon was always going to be a popular subject for film-makers to bring to life via the cartoon medium—but none can surpass Walt Disney Studios founded by Walt Disney himself for the sheer genius and cinematographic sorcery of its best productions.

The first dragon conjured up by Disney was Kenneth Grahame's rather fey specimen from his short story 'The Reluctant Dragon.' A 20-minute cartoon that remained relatively faithful to Grahame's tale (though St. George was replaced in the cartoon by St. Giles) was part of a full-length feature film also entitled *The Reluctant Dragon*. Released in 1941, this film centres around a live-action tour of the Disney studios by American wit and radio comedian Robert Benchley, where he is shown various complete cartoon shorts, ideas for future animated features, art and animation classes, plus much else. Although the film itself was never re-released in its full-length version, its 'Reluctant Dragon' segment was later released separately as a 'mini-classic' cartoon.

Originally, there were plans for the Jabberwock to appear in Disney's animated film *Alice in Wonderland* (1951), but its sequence was deleted before the film's release, together with its song, 'Beware the Jabberwock.' Illustrations taken from this unused sequence, however, later appeared in a Disney-published illustrated book of Lewis Carroll's poem 'Jabberwocky'—and, very incongruously, the supposedly ferocious dragon in question was depicted wearing a purple sweater!

In 1959, Disney's next major dragon set the big screen aflame, metaphorically if not literally. This was due to the spectacular animation featured in *Sleeping Beauty*, vividly portraying the transformation of the evil fairy Maleficent into a huge, fire-breathing dragon ablaze with incandescent, bat-winged fury, seeking to incinerate Prince Philip as he bravely strives to rescue Princess Aurora, the Sleeping Beauty of this film's title.

A much more comical, purple-haired dragon, yet just as eager to gobble up its antagonist—in this instance a somewhat scatter-brained Merlin the magician—is one of many forms assumed by the decidedly deranged witch Madam Mim in Disney's zany adaptation of T.H. White's popular Arthurian novel for children, *The Sword in the Stone* (1938). Disney's animated version was released in 1963.

The biggest dragon star from the Disney studios, however, was Elliott, from the live-action/animated musical feature film *Pete's Dragon*, first released in 1977. Elliott is a huge pot-bellied green dragon with a shock of pink hair and a pair of unfeasibly small wings (yet which nevertheless enable him somehow to become airborne should he need to be). However, he is generally visible only to his young owner, a small orphan boy named Pete. Elliott accompanies Pete, as his friend and protector, to the coastal fishing town of Passamaquoddy in Maine, USA, where the boy has fled in order to get away from his abusive adoptive hillbilly family, and the two soon become embroiled in all manner of amusing slapstick scrapes and general chaos.

One UK-released DVD edition of *Pete's Dragon* included as an extra feature a little-known Disney documentary entitled *Man, Monsters and Mysteries* (1974). This included a delightful animated segment featuring Nessie, the Loch Ness monster—considered by some to be a bona fide water dragon. Far removed from the dark, sleek, mysterious entity of cryptozoological fame, however, Disney's version is an affable multi-coloured beastie adorned with red polka-dots and voiced by Sterling Holloway. In 2011, Disney announced the production of a new cartoon short, 'The Ballad of Nessie', featuring a green and rather more dragonesque monster.

Just as Disney's celebrated animated film *Fantasia* (1940) consisted of a series of cartoon representations of famous pieces of classical music, *Musicana* was planned to be a comparable film showcasing via cartoons a series of folktales from around the world, backed once again by classical music. Sadly, this potentially spectacular film, whose development began during 1982-1983, was never produced. However, preparatory drawings and other preliminary artwork created for it still exist, giving a tantalising glimpse of what might have been. One particularly striking series of full-colour pastel artwork is from a sequence by artist Mel Shaw designed to illustrate Meso-American folklore, and includes beautiful renditions of Quetzalcoatl, the Mexican plumed serpent.

In 1997, the Disney animated feature film *Hercules* was released, and, befitting a movie based (albeit loosely) upon tales

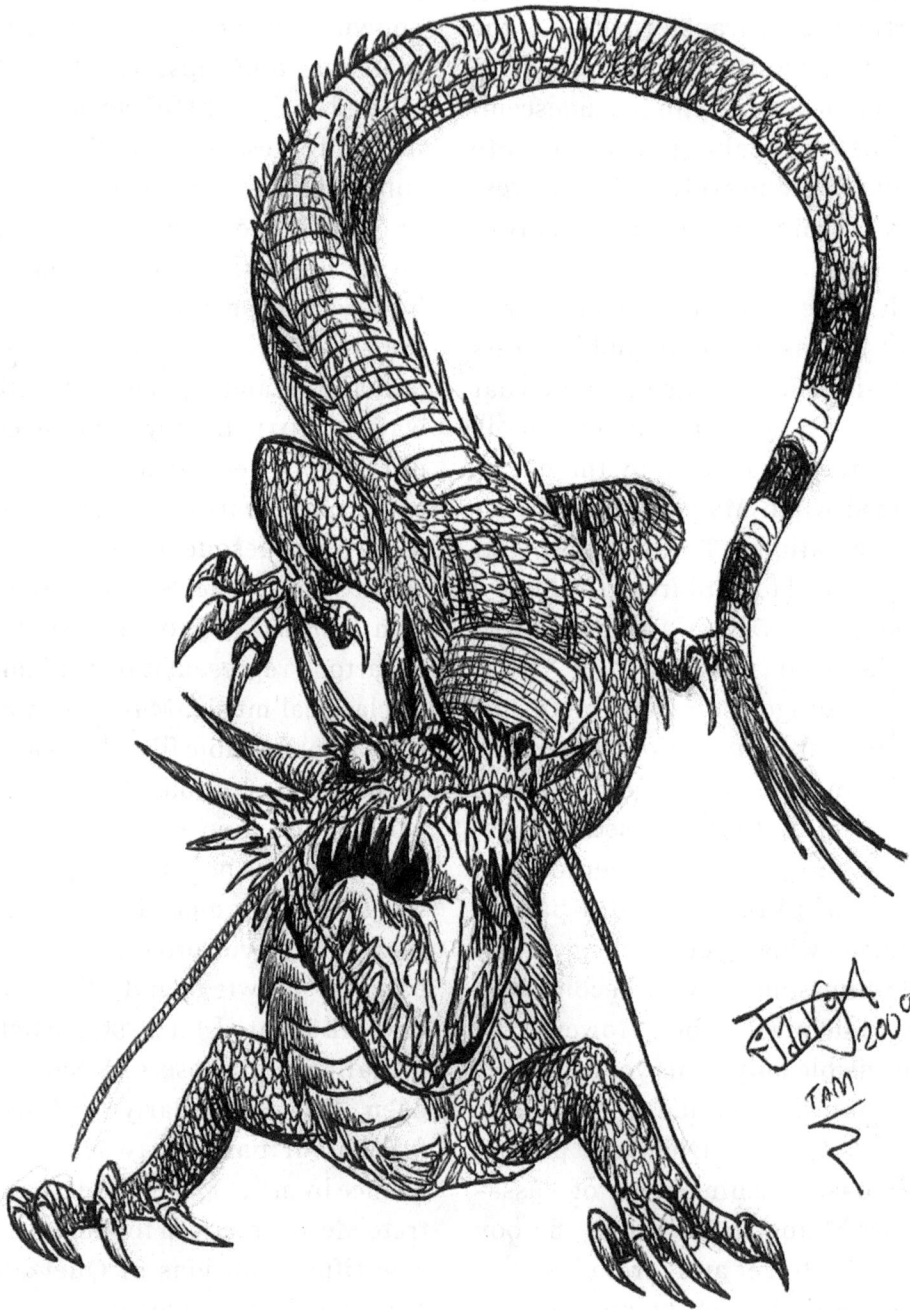

The dragon is nowadays a major movie star in its own right. (Hodari Nundu)

from Greek mythology, it included an epic battle between the young demi-god hero and the Lernean hydra. This multi-headed dragon has been summoned by Hades to destroy Hercules, but when he successfully kills it by causing a landslide, our hero finds himself elevated to celebrity status among the general public.

A year after *Hercules*, Disney released *Mulan*, a full-length animated musical that retold the Chinese legend of Hua Mulan, the daughter of elderly warrior Fa Zhou, who disguises herself as a male warrior in order to take her ailing father's place and battle an invading Hun army. Assisting Mulan in her endeavours is a small red Chinese dragon, Mushu, ostensibly her guardian but not overly brave. Voiced by American actor-comedian Eddie Murphy, Mushu is the film's principal comic-relief character, and reappears in the direct-to-video sequel, *Mulan II* (2004).

Dragons have appeared in big-screen cartoons made by other film production companies too, all over the world, but especially (as might be expected) in the Far East. For example, *Little Nezha Conquers The Dragon King* (1979) was a sumptuously-produced Chinese animated film that drew upon ancient Chinese mythology and followed the exploits of young warrior-deity Nezha. Thanks to his training by the immortal teacher and reincarnated Shang emperor Taiyi Zhenren, and after many trials and tribulations along the way, Nezha successfully defeats Ao Guang, the mighty Dragon King of the East Sea, thereby bringing peace to the Zhou Dynasty.

In 2002, the Academy Award or Oscar for Best Animated Feature was won by *Spirited Away*, a remarkable Japanese fantasy film written and directed by Hayao Miyazaki, in which a young girl, Chihiro Ogino, becomes trapped in a bizarre alternate reality populated by spirits and monsters while seeking her parents. One of the principal characters is Haku, a river spirit, who sometimes assumes the form of a young boy, but for much of the time appears as an enormous white Oriental dragon, who can fly without wings in the manner of many such dragons from the Far East.

In 1972 Australia released its first home-grown animated film musical, *Marco Polo Junior Versus The Red Dragon*. Voiced by American singer and former teen idol Bobby Rydell, Marco Polo Junior is the fourteenth heir of the famous Italian traveller whose name he shares, and journeys to the legendary Chinese kingdom of Xanadu to unite the two halves of a mystical medallion. While there, he rescues the beautiful Princess Shining Moon from a forced marriage, outwits a pair of bumbling spies, encounters a hypochondriac dinosaur, and confronts Xanadu's comically despotic ruler, the Red Dragon.

Vera Chapman's Arthurian novel *The King's Damosel* (1976) was the basis of an animated film entitled *Quest For Camelot*, produced by Warner Brothers and released in 1998. Telling the story of the quest for King Arthur's legendary sword Excalibur by reclusive blind youth Garrett and plucky teenage girl Kayley (whose bold ambition is to become a Round Table knight), it features for comic relief an amusing two-headed dragon called Cornwall (the uncouth head) and Devon (the sophisticated head). Voiced by Don Rickles and Eric Idle, its two heads ostensibly dislike each another, but ultimately come to realise that they are happier together than apart.

A number of 'made for television' cartoon films have featured dragons, but the most significant of these is *The Flight of Dragons* (1982), produced by Rankin/Bass. It was primarily inspired by Gordon R. Dickson's novel *The Dragon and the George* (1976) and its various *Dragon Knight* sequels, but was also influenced by Peter Dickinson's speculative natural history book *The Flight of Dragons* (1979) and its evocative illustrations by Wayne Anderson. The central theme of this very vivid, colourful film, filled with warring wizards as well as spectacular dragons of several different types and behaviour, is whether the worlds of magic and science can co-exist or whether one is destined to supplant the other.

LIVE-ACTION AND CGI
MOVIE DRAGONS

Creating a realistic live-action dragon on screen is clearly a more difficult task than simply drawing one for an animated film, but thanks to stop-motion special effects and the marvels of modern-day CGI (computer-generated imagery), some truly breathtaking successes have been achieved, which include the following examples.

The undisputed master of special effects achieved via the use of expertly-constructed

models in conjunction with painstaking stop-motion photography was the late Ray Harryhausen (1920-2013), who personally developed an extremely effective, advanced version known as Dynamation. His meticulous work in this field turned films such as *The Beast From 20,000 Fathoms* (1953), the Sinbad trilogy (1958, 1974, 1977), *Jason and the Argonauts* (1963), *One Million Years B. C.* (1966), *The Valley of Gwangi* (1969), and *Clash of the Titans* (1981) into cinematic masterpieces of fantasy and science-fiction.

Many famous mythological beasts featured in his fantasy films, such as a roc, griffin, centaur, harpies, cyclops, winged homunculus, and snake woman, as well as two different dragons. One of these latter was a guardian dragon that appeared in *The 7th Voyage of Sinbad* (1958). A typical wingless classical dragon that breathes fire, it is sent to kill Sinbad and his men by a villainous wizard called Sokurah. Fortunately, they are able to slay it using an enormous crossbow-like ballista, and as it falls, mortally wounded, it crushes Sokurah to death beneath its huge body—two enemies duly dispatched for the price of one!

In *Jason and the Argonauts* (1963), the Golden Fleece sought by Jason and his men is guarded by the Colchis dragon. Although this is usually depicted as a winged classical dragon, for maximum visual appeal Harryhausen represented it in the film as a multi-headed hydra-like version instead. It kills one of Jason's men, the treacherous Acastus, before being slain by Jason himself, who is then able to steal the Golden

Fleece, and later returns with it in triumph to Thessaly.

There is no absolute consensus as to whether Godzilla (aka Gojira) is meant to be a radioactivity-engendered mutant dinosaur, a giant amphibious lizard, a modern-day dragon, or a combination of all three, but there is no doubt that its arch-enemy King Ghidorah is a dragon—a limbless, two-winged, twin-tailed, triple-headed, golden-scaled, fire-vomiting, laser-spewing dragon, to be precise! Ever since Godzilla debuted in the 1954 Japanese film of the same name, he has faced a daunting array of monstrous antagonists, but King Ghidorah is the most impressive,

Greeted by Godzilla! (Dr. Karl Shuker)

and has appeared in five Godzilla films so far. The first was *Ghidorah, the Three-Headed Monster* (1964), in which it reaches Earth inside a meteorite from Outer Space, and proceeds to decimate Japan until finally sent packing by Godzilla, Rodan (a mutant pterosaur), and Mothra (a giant moth).

Dragons on the silver screen took an enormous step—or wingbeat—forward in 1981, with the release of *Dragonslayer*, a co-production between Walt Disney Studios and Paramount Pictures, which featured what was then the most realistic and visually stunning dragon ever seen in the movies. Bearing in mind that a quarter of

the film's entire budget was spent upon designing and breathing life into its reptilian star attraction—a fire-breathing winged classical dragon called Vermithrax Pejorative who is appeased only by being fed two virgin maidens each year until sorcerer's apprentice Galen sets out to destroy it and its nest of rapacious dragonets—it is little wonder that the results were so eye-popping. Several highly complex, multi-part models were created, including one of its head for close-ups, a flying model, and a walking model, thereby eliminating the need for stop-motion cinematography by using a new technique called go-motion, in which the model was moved slightly *while*

A dragon, inspired by *Q—the Winged Serpent*, by fantasy artist Anthony Wallis.

the camera was filming (rather than the camera filming a frame *after* the model had been moved).

One of the quirkiest dragon films ever released was director Larry Cohen's *Q— The Winged Serpent* (1982), in which a cult in New York City successfully resurrect the ancient flying serpent deity of Aztec mythology, who proceeds to swoop down from Manhattan's skies and skyscrapers to seize, dismember, and devour unwary city dwellers. In appearance, this odd-looking entity is not serpentine at all, instead resembling a rather gangly, long-necked quadrupedal dragon with wings, but its smooth skin seems devoid of typical reptilian scales or spines, and does not sport any feathers either (despite the original Quetzalcoatl being a plumed sky serpent).

By the mid-1990s, on-screen dragons had begun to go digital, as evinced by the CGI-created specimen voiced by Sean Connery in *Dragonheart* (1996). When Draco, the world's last dragon, encounters Bowen (played by Dennis Quaid), the world's last dragon-slayer, it initially appears that only one of them will survive their meeting. Happily, however, some enlightening conversation convinces them to join forces instead, and the scene is set for a dynamic confrontation with the evil King Einon.

Perhaps the epitome of the modern-day dragon film, however, is *Reign of Fire* (2002), directed by Rob Bowman, and starring Christian Bale and Matthew McConaughey. Set in the year 2020, this postapocalyptic film reveals the devastation that has resulted after a sleeping dragon was inadvertently chanced upon and woken in an underground cave during some construction work on the London Underground shortly after the beginning of the new millennium. The dragon forced its way to the surface, swiftly multiplied, and within a dozen years humanity was virtually wiped out by a worldwide plague of flying fire-breathing dragons. Finally, however, a brave survivor, Quinn Abercromby (played by Bale), and his isolated community hiding out in a Northumberland castle reluctantly join forces with a band of American fighters led by Denton Van Zan (McConaughey) to bring to a decisive end the dragons' literal reign of fire. Although the story's premise seemed decidedly far-fetched, the special effects were truly astonishing.

The same is true of *Dragon Wars* (2007), a South Korean film released in the West. In it, a benevolent imoogi and a malevolent imoogi (Korean serpent dragons) battle for supremacy, the latter employing an army of Western dragons, humanoid warriors, and dinosaurian monsters, and razing much of Los Angeles in the process.

An intriguing update of a classic story is the premise of the 2011 film *Age of the Dragons*. Here, Herman Melville's timeless novel *Moby Dick* is reinvented as a search by an alternate-world Captain Ahab (played by Danny Glover) and his crew not for a great white whale, but rather for a great white dragon.

So far, the dragons and dragon-riders of the planet Pern, chronicled in the extensive

Dragons preying upon humans have always
been the stuff of movies! (Thomas Finley)

series of novels by Anne McCaffrey, have not been portrayed on the big screen. However, an equally outstanding set of space dragons that *have* been portrayed are those of Pandora—a lush verdant moon orbiting an enormous gas giant planet, Polyphemus, in James Cameron's blockbuster film *Avatar* (2009).

One of the most dramatic Pandoran species is the ikran or mountain banshee. Somewhat pterodactylian in superficial appearance, this is a huge mountain-dwelling aerial carnivore with two pairs of leathery wings (larger fore and smaller aft, boasting an average wingspan of 45.5 ft for its fore pair). It can be ridden by only the bravest Na'vi warriors (the Na'vi are the 9-10-ft-tall blue-skinned humanoids native to Pandora). A smaller relative is the ikranay or forest banshee, inhabiting rainforests and sporting a 23-ft wingspan. Most spectacular of all, however, is the closely-related toruk or great leonopteryx, a brightly-coloured iridescent behemoth of the skies, which is the apex aerial predator of Pandora. Possessing a stupendous 75-ft-plus wingspan, it has even attacked human aircraft, believing them to be competing predators invading its territory. At the opposite end of the size scale is the riti or stingbat, a butterfly-like dragon indigenous to the rainforest canopy, with a wingspan of just 4 ft, and of only very limited intelligence. Although very aggressive and armed with lethal tail spines, these small creatures are treated almost as pets by some Na'vi, who feed them fruit by hand.

A rich variety of dragons have also featured in the 21st-Century screen versions of J.K. Rowling's *Harry Potter* novels, Tolkien's *Lord of the Rings* trilogy and the three-part *The Hobbit* (all directed by Peter Jackson), and Christopher Paolini's *Eragon* in his *Inheritance Cycle* series. All in all, there is every reason to believe that the age of the dragon will live on and attain even greater heights of awe-inspiring wonder in future generations of films on the big screen.

DRAGONS ON TELEVISION

There have been many notable small-screen dragons, but thanks to the charmed tenacity of nostalgia, perhaps those that we most readily recall are ones that featured in shows from our childhood.

One of the legendary names in British children's TV is Oliver Postgate (1925-2008), who created and wrote some of the most beloved shows of all time in this special genre of television—*Bagpuss*, *Clangers*, *Noggin the Nog*, *Ivor the Engine*, and *Pogles' Wood*, among others. They were all made by his company Smallfilms (founded with Peter Firman), and screened by the BBC. Some of these featured delightful dragons, remaining cherished childhood memories for generations.

Originally screened from 1969 to 1972 but repeated numerous times thereafter, *Clangers* was a stop-motion show of 27 10-minute episodes. They featured a family of small whistling aliens, the clangers, with long snouts and knitted waistcoats. These share a tiny hollow planet with a host of exotic fauna and flora, such as the iron chicken, the froglets, the musical trees, and, most notable of all, the soup dragon. It is she who obtains from the planet's volcanic soup wells the delicious blue string pudding and green soup that the clangers adore. It was this character (and her baby dragon) who inspired the name of Scottish alternative rock band The Soup Dragons.

Consisting of 27 10-minute episodes (six in colour) of limited stop-motion photography and first screened in 1959, *Noggin the Nog* was a Norse-type saga about a tribe of Northmen, the Nogs, led by King Noggin, and featuring an extensive cast of characters. These include Noggin's villainous uncle Nogbad the Bad, inventor Olaf the Lofty, a giant green bird called Graculus, Arup the great walrus, and an amiable ice dragon known as Groliffe (not to mention a flying machine and a fire machine!). Befriended by Noggin, Groliffe subsequently comes to his aid when he and his friends are in trouble.

Spanning 1959 to 1977 and consisting of 32 10-minute black-and-white episodes and 40 5-minute colour episodes of stop-motion photography, *Ivor the Engine* was famously set in "the top left-hand corner of Wales." It features a green locomotive called Ivor, his driver (Edwin) Jones the Steam, plus several supporting characters. Notable among them is Idris, a small red heraldic dragon based upon the emblem of Wales, who lives with his wife and two dragon children in an extinct volcano called Smoke Hill, and sings in the local choir.

A dragon called Dennis who combined the best of both geographical types appeared in *James the Cat*—a cartoon series of 52 5-minute episodes screened by the BBC from 1984 to 1992. One of many animal friends of the show's title character, Dennis is a pink Chinese dragon but breathes fire and speaks with a Welsh accent!

A happiness-bringing luck dragon long before Falkor debuted in the novel and film versions of *The Neverending Story*, Chorlton was the friendly but somewhat slow-witted star of an enchanting British series entitled *Chorlton and the Wheelies*, originally screened on ITV from 1976 to 1979. In the very first of its 40 stop-motion animated episodes, created by the company Cosgrove Hall, Chorlton hatches from an egg and then arrives in Wheelie World. This is a strange land populated mostly by Wheelies—creatures that have wheels instead of legs, but which are burdened with sadness conjured up by a wicked witch called Fenella . . . until Chorlton's happiness soon dispels the gloom. In subsequent episodes, Fenella puts into practice all manner of evil schemes to rid Wheelie World of Chorlton, or cause problems for him, but he and his Wheelie friends invariably manage to foil them.

One of the most popular series from the golden age of children's TV in the USA was *H. R. Pufnstuf*, a live-action show featuring life-sized puppets whose 17 25-minute episodes were first screened from September 1969 to September 1971. H. R. Pufnstuf is not only a dragon but also a mayor—of a magical isle called Living Island. Here everything is alive, even the houses, and is where an 11-year-old boy called Jimmy (played by Jack Wilde, the artful dodger in the 1968 film musical *Oliver!*) and his talking flute Freddy are taken to in a mysterious boat. The series' basic scenario is similar to that of *Chorlton and the Wheelies*, in that the bane of Living Island is a troublesome witch, called Witchiepoo here, but her evil plans are always thwarted by the dragon, Jimmy, and their many friends there.

Dragons in children's TV shows tend to be whimsical rather than wicked.

Originally aired in Canada and the USA from 1993 to 1997, and running to five seasons, collectively containing 65 30-minute episodes, *The Adventures of Dudley the Dragon* was a live-action show in which a full-costumed actor played Dudley. Befriended by two children after waking up from centuries of hibernation, Dudley finds out what the modern-day world is like, with particular emphasis upon environmental issues.

Other popular children's TV shows featuring dragons included *Wacky Races*, *My Little Pony*, *The Smurfs*, *Pocket Dragon Adventures*, *Eureeka's Castle*, *Digimon*, and, for older children and teenagers, *Power Rangers*, *Dungeons and Dragons*, and several Manga series. Moreover, countless TV cartoons have featured dragons as one-off foes or comic relief characters.

One of the earliest but most cherished television shows to include a dragon character was *Kukla, Fran, and Ollie*. Airing on American TV from 1947 to 1957, this puppet series was intended for children but proved more popular among adults, due to its being entirely ad-libbed. Its leading characters included Ollie, or, to give him his full name, Oliver J. Dragon—a one-toothed dragon with a very roguish persona.

Genuine adult programmes that contained dragons in their dramatis personae have been somewhat few and far between, but the following two are among the best known examples, though very different indeed from one another.

The Munsters was a very popular American sitcom of the mid-1960s, famous for its storyline of a family whose members are all bona fide monsters—a Frankensteinian head of the household, his vampire wife, another vampire as Grandpa, their young werewolf son, and—horror of horrors—a totally normal niece! (The show's running joke was that she was the freakish member of the family!) Among their equally bizarre pets was a dragon called Spot.

Merlin was a prime-time British fantasy show set in the Arthurian age, but when the magician Merlin was still a youth and his equally young friend was the headstrong and somewhat arrogant Prince (later King) Arthur. During the series, Merlin (played by Colin Morgan) learns a great deal of sorcery from Kilgharrah, the Great Dragon (voiced by John Hurt). He was an original character created specially for this show, who acts as mentor, protector, and advisor to the young wizard, nurturing and honing his developing magical skills. A deadly cockatrice also appears in one episode. The first series was screened by the BBC in 2008, and four more were subsequently produced and broadcast; the last episode of the final, fifth series was broadcast in two parts on Christmas Eve 2012.

Moving right up to date: the American blockbuster fantasy TV show *Game of Thrones* features dragons of the fire-breathing winged classical version, which exist in three different colour varieties. Until recently, they were thought to have become extinct more than 150 years ago. Although they cannot be tamed, dragons can be trained and mastered, and have been utilised and sometimes even ridden by humans in battle. The show is based upon the best-selling series of novels *A Song of Ice and Fire*, written by George R. R. Martin, and the first novel shares the TV show's title. Its fourth season is due to air in 2014.

JUST AS DRAGONS have featured extensively within the visual arts, so too have they appeared in substantial numbers and variety within our literature and music.

Cartoon dragons are perennially popular characters in children's television.

CHAPTER 7:
DRAGONS IN LITERATURE, MUSIC, SPORT, AND HOBBIES

WITH FLAMES just as brilliant and as searing as any that they have spurted forth at their challengers, dragons have fired the imaginations of many of the world's most celebrated writers, composers, and purveyors of culture.

DRAGONS IN CONTEMPORARY FICTION
When traditional myths and folktales preserved for countless generations by oral story-telling and by recycling in the works of the ancient Greek, Roman, Oriental, and other world scholars eventually gave way during medieval times to newer, more formal literature, many of the dragon's legendary opponents from those bygone ages fell by the literary wayside. In stark contrast, conversely, the dragon itself flourished—adapting and evolving as vigorously as ever, in order to seize its place in the modern-day medium of the printed word as a principal, multifaceted character that remains just as significant in today's fiction as it once was in the fables of yesteryear.

True, several significant works of medieval and early modern literature do feature dragons. These include *Beowulf*, the Volsung saga, *Sir Gawain and the Green Knight*, and Edmund Spenser's *The Faerie Queene* (1590). Moreover, many references and allusions to a wide variety of dragon types can be found in Shakespeare's plays

As might be expected, however, contemporary fantasy novels and short stories are positively replete with dragons of every conceivable—and inconceivable!—variety. There are gigantic dragons, like the 6560-ft-long individual in Lucius Shepard's short story 'The Man Who Painted the Dragon Griaule' (1984), and minuscule dragons, such as the tiny jheregs in Steven Brust's Vlad Taltos novels (1983 onwards) and the mute, minute specimens in Robin McKinley's *The Hero and the Crown* (1984). There are sentient, talking dragons, like those Norse-inspired versions in J. R. R. Tolkien's Middle-Earth, and the enormous flightless dragons of Venus with a penchant for scientific research that appear in Robert Heinlein's *Between Planets* (1951).

Confrontations between different types of dragon also occur, as in Ron Preiss and Michael Reaves's epic fantasy novel *Dragonworld* (1979), featuring a climactic

Not all dragons of literature are gigantic. (Thomas Finley)

battle between the world's last surviving dragon and a unique, embittered hybrid of dragon and cold-drake. There are pet dragons, such as the dog-sized swamp dragons in Terry Pratchett's *Discworld* series (1983 onwards). And there are frost dragons like Ingeloakastimizilian and shadow dragons

The White Worm, from Bram Stoker's novel *The Lair of the White Worm* (1911), as portrayed by Pamela Colman Smith in its first edition.

like Shimmergloom, both of which appear in R. A. Salvatore's *Forgotten Realms* series (1988 onwards). Even cannibalistic dragons sometimes occur, most notably in C. S. Lewis's *The Pilgrim's Regress* (1933), in which the cold Northern dragon recalls how he ate his wife, claiming that worms can only mature into dragons if they devour their own kind (echoing the famous Latin maxim 'Serpens, nisi serpentem comederit, non fit draco').

Venerable dragons in fantasy novels include Mayland Long, now a man but formerly a Chinese dragon, appearing in R. A. MacAvoy's *Tea With the Black Dragon* (1983); and accursed dragons include Belial, originally an angel, who has transformed himself into an immense, evil, and totally insane dragon in Steven Brust's *To Reign in Hell* (1984). There is even an avowed Marxist dragon, Falameezar-aziz-Sulmonmee, occurring in Alan Dean Foster's *Spellsinger* series (1983-1994).

In Robin Hobb's *Realm of the Elderlings* series (1995), the elderlings are human-dragon hybrids, and there are even some living dragon statues. Shape-shifting dragons are commonplace too, like Villentretenmerth, the golden dragon from Andrzej Sapkowski's *The Witcher* series (1992); and Lady Arabella March, the titular worm in Bram Stoker's classic horror fantasy *The Lair of the White Worm* (1911). Entire societies of telepathic dragons also exist, such as the Kantri in Elizabeth Kerner's *Song in the Silence* (1997). And alien space dragons secretly inhabit planet Earth, like those representing the dragon empire Draconizica and residing on our planet since Precambrian times, as revealed in Nipaporn Baldwin's recently-published *The Society On Da Run* series. Nor should we forget the fantasy genre's countless guardian dragons, mechanical dragons, invisible dragons, comical dragons, and many more.

DRAGONS OF MIDDLE-EARTH

Considered by many readers and literary critics alike as the most significant and influential modern-day works of fantasy fiction ever published, *The Hobbit* (1937), *The Lord of the Rings* trilogy (1954, 1954, 1955), *The Silmarillion* (1977), and J. R. R. Tolkien's other novels contain a number of important dragons.

It has often been supposed that his works are Christian allegories or are at least derived from Christian themes, and therefore comparable in terms of inspiration to the Narnia novels of C. S. Lewis. In reality, however, Tolkien's principal muse was the *Elder Edda*—a collection of Old Norse myths and legends preserved principally within the *Codex Regius*, which is a medieval Icelandic manuscript, written in the 13th Century. Yet the myths and legends themselves are far older, and include all of the famous Norse ones known today.

Many familiar character names in Tolkien's books, for example, including Gandalf and various of the dwarves featuring in *The Hobbit*, were borrowed directly by him from the *Elder Edda*. So too were other entities and themes, one of which that

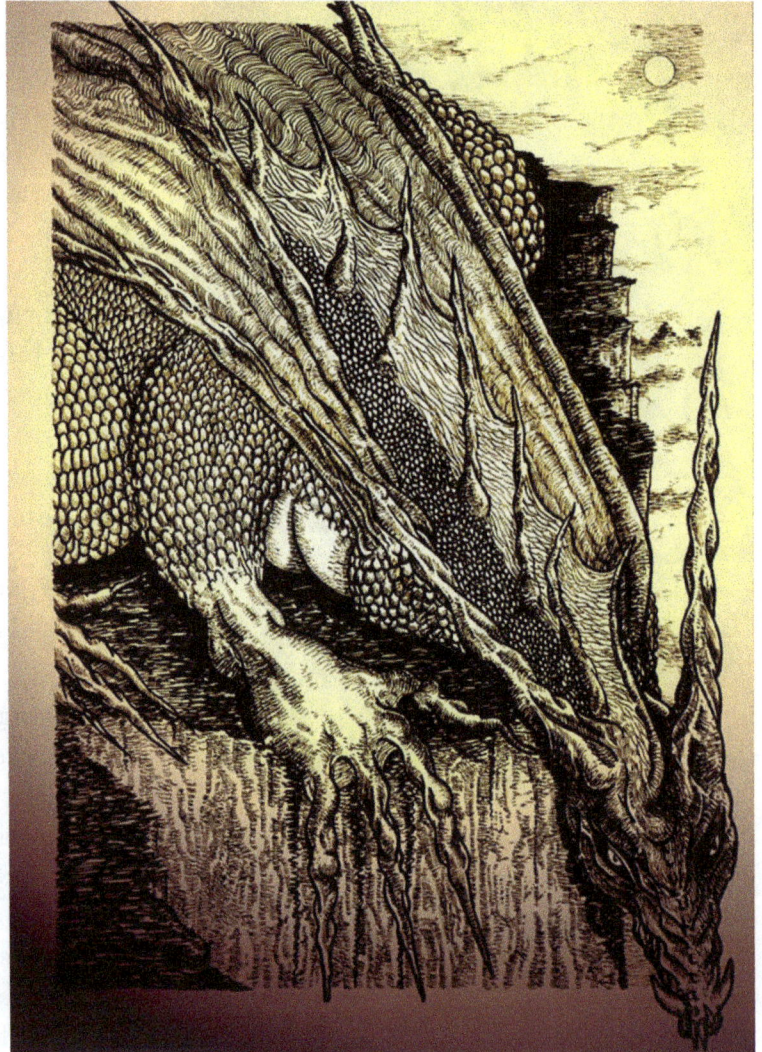

Tolkien's dragons were traditional Nordic forms, like this example from Andy Paciorek's book *Strange Lands*. (Andy Paciorek)

attracted his particular attention being the slaying by the hero Siegfried of the evil dwarf Fafnir, who had transformed himself into a typical Nordic dragon in order to protect his ill-gotten treasure hoard. This dragon became the basic template for the various examples featuring in Tolkien's novels.

Yet although Tolkien's dragons were of traditional Nordic form and treasure-hoarding behaviour, they were much more intelligent than their antecedents in the Old Norse myths and legends, and they could speak too.

Created by the dark lord Morgoth, there were three types—the great worms, the winged quadrupedal dragons, and the wingless quadrupedal dragons. Some could also breathe fire, enabling them to destroy the lands and cities of Middle-Earth's men, elves, and dwarves.

The first Middle-Earth dragon was Glaurung, a huge wingless fire-breather, which ultimately spawned numerous lesser dragons, and led them into battle on the side of Morgoth against the elves, but was finally slain by the hero-warrior Túrin.

The greatest Middle-Earth dragon of all, however, was Ancalagon the Black, the first winged fire-breather. His appearance alongside his spawn astonished the entire world, and initially gave victory to Morgoth's horde—until the great eagles and other warrior birds rallied against them, eventually achieving victory over their reptilian foes and breaking Morgoth's power forever. Ancalagon was slain by warrior Eärendil, and so enormous was this mighty

dragon's body that when it plummeted down from the sky to earth, it decimated the three-peaked mountain Thangorodrim.

The last famous Middle-Earth dragon was Smaug, a huge golden-red winged monster almost entirely ensheathed in impermeable iron scales. Smaug destroyed with fire the human city of Dale and vanquished the dwarves from the Kingdom under the Mountain (of Erebor) in order to seize for himself their vast treasure of gold, jewels, and precious elvish metals stored there. This he jealously guarded for almost two centuries until disturbed by a party of dwarves seeking retribution, led by Thorin Oakenshield and also including among their number the hobbit Bilbo Baggins. It was Bilbo who discovered the one vulnerable region on Smaug's underside, which enabled a Northman archer named Bard the Bowman to slay Smaug after he had attacked the city of Esgaroth upon the Long Lake.

More light-hearted and whimsical was another Tolkien dragon—Chrysophylax Dives. Pompously comical but still wily and villainous, he was finally captured and controlled by Farmer Giles of Ham in the book of the same title, using a mighty sword called Caudimordax (aka Tailbiter) that once belonged to a famous dragon-slayer.

THE DRAGONS OF PERN

One of the most intriguing series of dragon-based works of fiction comprises the *Dragonriders of Pern* novels by Anne McCaffrey, inasmuch as they very deftly combine and intermingle the ostensibly discrete genres of fantasy and science fiction. Set in a pre-industrial feudal society inhabiting a colony planet called Pern, the novels feature the tribulations brought upon the colonists by Thread. This is a deadly strain of fungal spore that periodically reigns down upon Pern from a rogue orbiting planet, the Red Star, killing everything that it touches and which, if not destroyed, would burrow into the earth and multiply rapidly. It was the devastation caused by Thread when the colonists first reached Pern that swiftly reduced their civilisation to its low level of technology.

To ensure that this pernicious organism never did take root, however, the colonists learnt how to tame a native species of fire-generating lizard, which, via genetic modification techniques, they then transformed into a much bigger, telepathic dragonesque species with great wings, which they now ride like scaly steeds through the sky, killing Thread with these neo-dragons' scorching fire. At the moment of its hatching from its egg, a dragon is psychically impressed upon a dragonrider, and remains so bonded for life. There are many different colours of dragon, related to their size. The largest is the bronze variety, and the largest bronzes are the queens, with which only the bronze males can mate.

The first of McCaffrey's Pern novels was *Dragonflight* (1968), which set the scene for more than 20 others, including *The White Dragon* (1978). This award-winning novel concerns a white runt dragon called Ruth, who demonstrates that size isn't everything by proving to be an exceptional

dragon even by the high standards typically set by Pern's winged wonders.

MAN-MADE FANTASY DRAGONS

Not all dragons in fantasy novels are of the straightforward flesh-and-blood variety. In *Havemercy* (2008), by Jaida Jones and Danielle Bennett, the dragons are sentient, magically-animated, metallic entities piloted by the renegade airmen constituting the elite Dragon Corps that have enabled the country of Volstov to achieve almost total defeat of its neighbouring enemy, the Ke-han Empire, during an ongoing war spanning more than a century.

In 'The Dragon' (1955), a short story interweaving medieval times and the present day, Ray Bradbury tells of how two knights are sent to destroy a terrifying cyclops dragon—exhaling great gusts of scorching air, plated all over in metal, and possessing just a single brightly-glowing eye set in the centre of its head. Despite charging at this formidable monster bravely, both knights are swiftly cut down and killed by it—and little wonder, for the 'dragon' turns out to be a modern-day steam engine, with its eye being the train's head light!

In Edith Nesbit's *The Last of the Dragons* (1925), the petrol-drinking dragon in question is so bored by constant expectations that it should battle princes that it finally persuades the king to transform it into a machine instead, and becomes the world's first aeroplane!

Speaking of which: in Michael Swanwick's novel *The Iron Dragon's Daughter*

(1993), a faerie slave girl called Jane successfully escapes from a dragon factory, where she helps to construct part-magical part-cybernetic dragons used as jet fighters, thanks to the assistance of an ancient rusted dragon called Melanchthon. A sequel, *The Dragons of Babel*, was published in 2008.

And in Chris d'Lacey's series of novels *The Last Dragon Chronicles* (2001-2012), long after the true corporeal dragons have seemingly died out college student David Rain is amazed to discover that the house where he lodges contains several living dragons composed of clay, created by the house's owner, potter Liz Pennykettle.

DRAGONS IN CHILDREN'S FICTION

The nature and roles of dragons that appear in children's fiction are directly influenced by the age-group at which the books are aimed. Those written for older children and teenagers often feature dragons that are much the same as those in adult fiction— ferocious, fire-breathing, mostly malevolent entities, to be battled and overcome by the heroes.

Such dragons also occur in the Brothers Grimm's fairy tales, because these were originally compiled for adults, not children. Sometimes, however, the dragon itself is portrayed as heroic, a noble companion assisting the human hero to conquer the enemy.

One of the most famous recent examples of the latter scenario is Christopher Paolini's four-novel epic fantasy series *The Inheritance Cycle* (2002-2011). It began with the initially self-published bestseller *Eragon*, in which the eponymous teenage dragon-rider and his ever-loyal dragon Saphira seek to overthrow Galbatorix, the evil monarch of Alagaësia, who was once a dragon-rider himself until his dragon was slain, leaving him so traumatised that he became insane.

Highly intelligent dragons devoted to their owners also feature extensively in Jane Yolen's *Pit Dragon* series (1982-2009), set in the far-distant future on Austar IV, a desert planet. But here they are bred to fight other dragons in a huge pit for sport.

Very different, yet still portraying dragons in a positive light, is Peter Dickinson's *The Tears of the Salamander* (2003). It tells of Alfredo's remarkable power to speak to the singing, fire-embracing salamanders inhabiting Italy's Mount Etna, whose precious tears can heal any wound or illness, and how he and they combine forces to conquer the menacing terror of his mysterious Uncle Giorgio, the evil, monstrous Master of the Mountain.

Dragons of a more traditional, ferocious nature appear in various volumes of *The Spiderwick Chronicles* (2003-2004) and *Beyond the Spiderwick Chronicles* (2007-2009) by Tony DiTerlizzi and Holly Black. These include serpent dragons like the venomous worms reared by the evil ogre Mulgarath, and a huge many-headed hydra with gills known as the Wyrm King. Malign dragons in dire need of slaying cause all manner of problems for Simon St. George in Jason Hightman's *The Saint of Dragons*

Marauding dragons were a perennial problem generally
requiring the services of a valiant dragon-slayer
in Grimms' fairy tales. (Thomas Finley)

series (2004-2006), due to his unexpected discovery that he shares not just his name but also his bloodline with the most famous of all dragon-killers, St. George. And dragons are even more intimately in the blood of Sam, in Thomas Bloor's *Worm in the Blood* (2005), because he makes the horrifying discovery that he is gradually transforming into a dragon, and that this curse, originating long ago in the Far East, has afflicted his family for generations.

J. K. Rowling's *Harry Potter* series (1997-2007) is amply supplied with dragons of many different types, which are often more reminiscent name-wise of dog breeds than dragon species—Norwegian ridgeback, Swedish short-snout, and Hungarian horntail, to name but three. They are entirely animal in form and behaviour, with no human attributes. Also of note here is the deadly basilisk inhabiting the Chamber of Secrets, which, unusually for this dragon type, is of monstrous size. Moreover, in stark contrast to traditional legends and folklore appertaining to basilisks, not only does it kill outright anyone who directly meets its lethal gaze but it also causes anyone who only looks at it indirectly, via a mirror, to enter a state of stony petrification.

In Ursula LeGuin's *Earthsea* trilogy (1968-1972), the two opposing scenarios of the dragon as evil destroyer and as noble benevolent are deftly blended. The Earthsea dragons begin as typical malign reptiles, but as the series progresses they evolve into more refined, regal entities that ultimately become near-deities, who in some cases can transform into humans—with which, the reader eventually discovers, they share a common ancestry.

In books for younger children, conversely, more often than not the dragons are gentler, quirkier, even comical creatures, their role being one of providing kindly aid and friendship to the child characters rather than confrontation, or even becoming mischievous pets, causing all sorts of amusing havoc. And even if they are unpleasant entities, they are usually too dim-witted to cause real problems, and can be readily thwarted by a plucky youngster.

One of the earliest, best-known examples in this category is Kenneth Grahame's short story 'The Reluctant Dragon' (1898), originally included in his book *Dreams Days*. It tells of a timid dragon and the hero St. George, sent by the local townsfolk to dispatch it, both of whom are more interested in poetry than anything so uncouth and potentially violent as battling with one another. A similarly diffident dragon is Custard, owned by a little girl called Belinda, and appearing in Ogden Nash's poem 'The Tale of Custard the Dragon' (1936). Happily, however, Custard shows his true mettle when Belinda and her animal friends are threatened by a peg-legged pirate.

In *My Friend Mr Leakey* (1937), a much-loved children's novel written by J. B. S. Haldane (who was also a famous British geneticist), the titular magician has a very exotic household. Its members include a jinn that runs errands, an octopus servant called Oliver, a tiny green cow to

'The Dragon Slammer' (Richard Svensson)

provide milk, and a small dragon called Pompey who lives amid the hearth fire's blazing coals, wears asbestos boots, and can be quite naughty at times.

Rosemary Manning wrote a trilogy of very charming dragon books for children—*Green Smoke* (1957), *The Dragon's Quest* (1959), and *Dragon In Danger* (1961). All three star a lucky green dragon who always refers to himself rather formally as R. Dragon. He is 1500 years old, displays impeccable manners at all times, has a great weakness for almond buns, loves retelling old Cornish and Arthurian folk stories, and never eats people as he considers such activity very impolite. His one human friend is Sue, a little girl who had chanced upon him while holidaying in Cornwall, and they soon become embroiled in all sorts of exciting adventures together.

In William Mayne's *The Worm in the Well* (2002), conversely, the titular worm is more than content to gulp down humans. And, in stark contrast to the typical worm's limbless form, this specimen has a rare abundance of legs, adding even more pairs each time it devours someone.

Michael Ende's bestseller *The Neverending Story* (1979) features a flying Oriental-style luck dragon called Falkor. He accompanies the boy-warrior Atreyu on his quest to discover a cure for the ailing Childlike Empress in the magical land of Fantastica.

Similar dragons, but this time of a terrestrial nature, feature in a delightful children's novel by C. S. Forester (better known for his stirring stories of naval warfare for

adults), entitled *Poo-Poo and the Dragons* (1942). It was originally conceived by Forester to induce his young son, suffering from food allergies, to eat, by reciting the story to him only if he ate his meals.

Sir John Tenniel's famous illustration of the Jabberwock for Lewis Carroll's classic children's novel *Through the Looking-Glass.*

In this book, a boy called Poo-Poo initially tries to hide from his father an extremely playful, multicoloured, and not-inconsiderably-proportioned dragon—"quite a fair size as dragons go—something between a duck and a motor bus"—that he happens to find on a vacant lot near his home. Later, however, the dragon brings home a friend, and the two dragons behave throughout the novel much like frisky, accident-prone pet dogs, albeit talking dogs, much to the delight of Poo-Poo and the frequent despair of his long-suffering parents.

In an interesting reversal of roles, Dick King-Smith's *Dragon Boy* (1993) tells the story of John Little, a young medieval human boy whose adoptive parents are dragons (albeit very anthropomorphic ones, named Montague and Albertina Bunsen-Burner!), because they have reared him ever since Montague found him as an abandoned orphan.

An unusually aggressive dragon to occur (albeit only briefly) in a book originally written for a young child is the Jabberwock. It appears in the poem 'Jabberwocky', which is contained in Lewis Carroll's immortal 'Alice' sequel, *Through the Looking-Glass* (1871).

Nor should we forget another equally ferocious, strange-sounding dragon found only within the realms of children's poetry—the dreaded hippocrump. Indeed, this is just one of a veritable menagerie of made-up monsters featuring in a wonderful volume of poetry by James Reeves entitled *Prefabulous Animiles* (1957); a sequel volume, *More Prefabulous Animiles*, was published in 1975.

DRAGONS IN COMIC-BOOK FICTION
Even the rarefied domains of comic-book superheroes are not entirely without a representation of dragons.

Fin Fang Foom, for instance, is a Marvel Comics creation who is originally awakened from his tomb by a teenager whose homeland is under threat by invaders from communist China—duly repelled by this humanoid dragon, who subsequently appears in many other adventures. Moreover, he is revealed to be an alien dragon entity from the world of Kakaranathara (aka Maklu IV) in the Maklu system of the Greater Magellanic Cloud.

Another Marvel Comics dragon of extraterrestrial origin, albeit occupying a lesser role than Fin Fang Foom, is Lockheed. Cat-sized and purple-scaled, he is the long-standing pet/companion of Shadowcat (the super-hero alias of Kitty Pryde), who is one of the X-Men.

Built by Prof. Gregson Gilbert of Empire State University and brought to life by alchemist Diablo, Dragon Man is an artificial dragon-like humanoid android and erstwhile supervillain in the Marvel Comics Universe. First appearing in issue #35 of Fantastic Four, he was created by veteran comic artists Stan Lee and Jack Kirby.

One of the most powerful and dramatic adversaries of DC Comics Universe superhero Green Arrow (aka Connor Hawke) is the Earth Dragon. Resurrected during an archery tournament, this ancient primordial fire-breathing female dragon is the physical personification of power itself.

The Japanese manga comic *Dragon Drive* by Kenichi Sakura was serialised in 14 volumes within the manga magazine *Monthly Shonen Jump*, ending on 5 January 2006. It features a number of dragons, partnered with human players in the eponymous virtual reality game, including a small lazy individual named Chibisuke, who proves to be the rarest of all dragons in the game and ultimately finds his true strength.

DRAGONS IN SPOOF AND SPECULATIVE FICTION

Dragons are popular subjects for spoof fiction—large, usually lavishly-illustrated books often purporting to be works of arcane natural history, but which upon reading are swiftly recognised as adroitly-constructed fiction penned with tongue very firmly in cheek.

An excellent example is *The Discovery of Dragons* (1996), written and sumptuously illustrated by Graeme Base. It reveals how Viking explorer Bjorn of Bromme, teenager Soong Mei Ying (daughter of a 13th-Century Chinese silk-trader), and 19th-Century cartographer-herpetologist Dr. E. F. Liebermann collectively discovered every known species of European, Asiatic, and tropical dragon—or didn't, as the case may be. An emphatic clue as to how sizeable the pinch of salt should be that the reader takes when perusing this sumptuous work appears on the front flyleaf, its blurb ending with the following unequivocal carpe diem:

> Other writers may be content to fall back on the old formula of painstaking research and careful documentation, but the internationally acclaimed creator of *this* splendid volume has adopted the far more efficient method of making it all up.

So don't say that you weren't warned!

Inventorum Natura: The Expedition Journal of Pliny the Elder (1979) is a spectacular tome compiled and exquisitely illustrated by fantasy writer-artist Una Woodruff. The premise behind this very skilfully-prepared volume is that it is a painstaking reconstruction of a supposedly long-lost work written in Latin by real-life Roman author-naturalist Pliny the Elder (23-79 AD), describing the astonishing fauna and flora that he allegedly observed during a purported three-year expedition to distant lands, an incomplete version of which Woodruff happened to rediscover. It includes several types of dragon—the pyrallis, basilisk, sea dragons, dragon-fishes, amphisbaena, Eastern dragons, a British hydra, and Western dragons.

Michael Green's *The Book of the Dragon-tooth* (1994) presents the transcribed journals of Magnalucius, a 16th-Century artist-mystic who sought the Perilous Dragon-tooth in order to tame this dragon relic's restless malice. *The Dragon Chronicles* (2002) by Malcolm Sanders claims to be the rediscovered 1500-year-old journals of the Great Wizard, Septimus Agorius, in which he records his epic journey through the kingdoms of Voorn, where he encountered a number of very formidable dragons.

Edited by Dugald A. Steer and profusely illustrated throughout, *Drake's Comprehensive Compendium of Dragonology* (2009) is purportedly the life's work of aptly-named Victorian dragon expert Dr. Ernest Drake. A veritable field-guide to the world's numerous types of dragon, it

American fantasy artist Thomas Finley's own delightful
interpretation of a humming-dragon. (Thomas Finley)

includes many so esoteric in form—such as the Argentinian humming-dragon—that they are totally unknown outside this tome's sumptuous covers!

Much smaller but very useful for slipping into one's pocket when setting forth in search of dragons is Lori Summers's *The Dragon Hunter's Handbook* (2002).

Pamela Wharton Blanpied's meticulously-conceived book *Dragons: The Modern Infestation* (1980), whose title refers to the outbreak of modern-day dragon sightings, is perhaps the most sober and quasi-scientific of all dragon spoof works, comprehensively documenting every aspect of dragon biology and behaviour. She even coins an official term, 'verminology,' for the scientific study of dragons, and although 'vermin' is a word normally associated today with rats and certain other rodents, it was originally derived from 'vermes,' which is Latin for 'worm.'

Rather more whimsical is Paul and Karin Johnsgard's delightful book *A Natural History of Dragons and Unicorns* (1982)—a subtle blend of genuine folklore, novel interpretations, and deftly-disguised cod science.

An excellent children's spoof book is Joseph Nigg's *How to Raise and Keep a Dragon* (2006), supposedly authored by John Topsell, distant relative of real-life bestiary writer Edward Topsell. This fun manual for anyone wanting a pet dragon is packed with information on dragon types, characteristics, and natural history, plus details of the equipment and supplies that every dragon owner needs (including an effective fire extinguisher!).

A very recent but also very satisfying volume is *A Natural History of Dragons: A Memoir by Lady Trent* (2013), compiled by Marie Brennan. This ornate tome records how eccentric naturalist Isabella, Lady Trent, became the world's pre-eminent dragon discoverer.

Most famous of all in this exceedingly rarefied literary genre is Peter Dickinson's elegantly-conceived, heady concoction of fact and fantasy, *The Flight of Dragons* (1979), lavishly illustrated by Wayne Anderson. In his captivating tome of 'faction', Dickinson theorises how dragons could have truly existed. Suggesting that they may have descended from dinosaurs, he even speculates upon how they could have evolved wings from their ribs in order to become airborne, and the highly-specialised hydrogen-producing physiology needed for them to have generated and expelled fire, which in turn would explain why there are no dragon fossils.

What makes *The Flight of Dragons* substantially different from the other titles discussed here, however, is that whereas those are all unquestionably spoof works, this book is better described as speculative fiction. In other words, one in which a wholly inconceivable prospect—that dragons in all their winged, fire-breathing, zoological absurdity constitute a genuine albeit now-extinct species—is examined and reasoned out in a no less rigorous and scientific manner than if it were a legitimate, realistic subject for deliberation.

DRAGONS IN SONG AND DANCE

Although the very considerable presence of dragons in contemporary fiction definitely exceeds that in music, they are far from unrepresented in this latter medium. As revealed here, dragons appear in several major works of classical music, and are also the theme of a number of memorable rock and pop songs. Moreover, the famous Dragon Dance provides an exuberant showcase for terpsichorean skill and spectacular design when celebrating the start of the Chinese New Year in major cities worldwide.

DRAGONS IN CLASSICAL MUSIC

Undoubtedly the most famous work of classical music to feature a dragon in it is Richard Wagner's *Ring Cycle* (*The Ring of the Nibelungen*), whose first complete performance took place in 1876. Inspired by the Volsung Saga—a 13th-Century work of Icelandic prose interweaving the Norse deities with the origin and decline of the Völsung clan (including Siegfried aka Sigurd and Brünhilde aka Brynhildr)—it consists of four linked operas. In the third of these, *Siegfried*, the eponymous hero prince slays Fafner (aka Fafnir), an evil treasure-hoarding giant (rather than a

dwarf, as in the original saga) who has transformed into a huge dragon in order to ensure that noone steals his treasure—which he has acquired by murdering his own brother for it (in the original saga, it was his father whom he killed).

Premiered in 1791, *The Magic Flute* was Wolfgang Amadeus Mozart's last opera, but also his most phantasmagorical, full of strange characters, arcane symbolism, and

Siegfried slays Fafnir, as depicted by Arthur Rackham.

even a fire-breathing serpent dragon. This formidable creature appears at the beginning of the opera, pursuing the somewhat less-than-heroic Prince Tamino, who promptly faints, and is only rescued when three female attendants of the Queen of the Night come to his aid and dispatch the vile creature for him.

Dragons and heavy metal music are a popular and potent combination. (Andy Paciorek)

A much more recent opera is *Grendel*, which premiered in Los Angeles in 2006. Composed by Elliot Goldenthal, it tells the medieval Anglo-Saxon story of Beowulf, but from the monster Grendel's point of view. It features the unnamed dragon from that story, who, in the opera's version, is consulted by Grendel for advice on how to become even more terrifying than he is already.

The Eastern dragon is represented metaphorically by the stirring Oscar-winning musical score written by Tan Dun for director Ang Lee's celebrated martial arts movie epic *Crouching Tiger Hidden Dragon* (2000). The score was later adapted by its composer to yield a classical concerto for cello and orchestra. The film title's 'crouching tiger' and 'hidden dragon' refer to the hidden talents and mysteries concealed within each person.

Peaceful Dance of Two Dragons, a powerful composition by Japanese composer Isotaro Sugata (1907-1952), blends traditional Japanese imperial music with several different Western influences. It features both an orchestra and traditional Japanese percussion.

DRAGONS IN POP AND ROCK MUSIC
The premier pop song concerning a dragon is, without any doubt, 'Puff the Magic Dragon'. Written originally as a poem by Leonard Lipton, it was later converted into a song by Peter Yarrow, and recorded in 1962 by the folk trio Peter, Paul and Mary (Peter being the selfsame Peter Yarrow). Inspired by Ogden Nash's poem 'The Tale of Custard the Dragon' and becoming a huge hit worldwide, the song tells the poignant tale of a lonely boy called Jackie Paper who makes friends with an immortal dragon named Puff in the magical, imaginary

land of Honalee. They have lots of adventures together until, one day, Jackie stops visiting Honalee to play with Puff, because he is no longer a boy—he has grown too old to believe in dragons and enchanted lands any more, leaving poor Puff to weep a sad tear for the loss of his friend.

Such was this song's continuing appeal that in 1978 a 30-minute animated TV film of the same title was released, whose plot was adapted from the song's lyrics. Nominated for an Emmy award, it was followed by two sequels. Burgess Meredith voiced Puff in all three films.

The animated film *The Flight of Dragons* (1982) features a lilting, wistful title song composed and sung by celebrated American singer-songwriter Don McLean, which complements the film's sometimes mystical ambience. Back in 1867, English actor-singer C. M. Leumane wrote the jaunty multi-verse song 'Whisht! Lads' ('Be Quiet! Boys'). Traditionally sung in Sunderland's Meckem dialect, it retells the famous northern folktale of the Lambton worm.

Due to its dark, ferocious image, the Western dragon features directly or as a metaphor in numerous songs (and even band names) within the heavy metal (HM) global genre of rock music, particularly in the sub-genres of thrash metal, power metal, symphonic metal, and progressive metal. Among the most notable examples are German thrash metal band Sodom's 'Magic Dragon' (1989), Swedish power metal band HammerFall's 'The Dragon Lies Bleeding' (1997), American HM band Nevermore's 'The River Dragon Has Come'

(2000), English power metal band Dragon-Force's 'Heart of a Dragon' (2003), French HM band Gojira's 'Where Dragons Dwell' (2005), and Italian symphonic power metal band Ancient Bards' 'Daltor the Dragonhunter' (2010).

'The Consummate Dragon,' based upon Smaug in Tolkien's *The Hobbit*, appears on American hardcore punk band Shai Hulud's second studio album *That Within Blood Ill-Tempered* (2003). 'Dragon' is a song from German heavy/power metal band Grave Digger's eleventh studio album *Rheingold* (2003), which was inspired by Richard Wagner's classical *Ring Cycle*. Several dragon songs are contained in Italian symphonic power metal band Rhapsody of Fire's five-album *Emerald Sword Saga* (1997-2004), which was directly influenced by the realms of high fantasy from the 'swords and sorcery' literary genre.

CHINESE DRAGON DANCE

Dragons appear not only in song but also in dance, and very prominently so in the Chinese Dragon Dance. This is performed in a lavish, skilfully-executed manner in New Year celebrations wherever there is a Chinese community.

The dragon itself comprises a long serpentine body, usually consisting of a lengthy series of hoops made from aluminium, plastic, bamboo, or very light wood, and covered in brightly-coloured fabric, with an ornamental head at one end and a tail at the other. The body is raised up by poles attached to the hoops, and can be 60-230 ft long, usually consisting of

9-15 sections but sometimes up to 46 in total. A sizeable team (often as many as 50 persons) is thus required to carry the dragon by raising the poles and to execute the complex movements constituting the dragon dance. The longer the dragon, the more good fortune it is believed to bring, so the longest dragons that are feasibly possible to construct and manoeuvre are utilised in major parades and ceremonies.

A dragon from a Chinese dragon dance.

Although rarely seen in Western displays, a highly complex double dragon dance is sometimes staged, featuring two troupes of dancers. Most intricate of all, however, is the incredible nine-dragon dance, in which nine separate dragon troupes interact together in a spectacular celebration of the number nine—an exceedingly lucky, 'perfect' number in Eastern tradition.

DRAGONS IN CONTEMPORARY SOCIETY
Today, the dragon is everywhere. On a hitherto-unparalleled scale, culture has indeed cultivated, and inculcated, this reptilian monster into the very fabric of our existence. Quite simply, the dragon is a modern-day megastar! From fashion to flags, collectables to computer games, kites to kimonos, netsuke to New Age, there are images of snarling, coiling, conflagrating dragons staring out at us from every conceivable type of media, not just books, films, television, and artwork.

DRAGONS IN FASHION
The dragon first became a fashion icon back in China's Yuan Dynasty (1271-1368 AD). This was when the five-clawed Chinese dragon was officially adopted as the emperor's own personal symbol. No-one else was allowed to wear garments bearing its image, or dyed golden-yellow in colour, upon pain of death. The four-clawed dragon was the emblem of nobility, whereas the three-clawed dragon was the emblem of the general populace.

Not only would anyone wearing a gold five-clawed dragon motif be executed, so too would every member of their clan—a law that perpetuated into the Ming Dynasty (1368-1644 AD). Little wonder, therefore, that this image became such a coveted emblem in later ages. Today, conversely, Oriental dragons of every shade and toe number, as well as Western dragons, are commonplace on a vast range of clothing, especially among the New Age and Goth cultures—everything from ornamental kimonos and dressing gowns to shirts, t-shirts, jackets, hoodies, elaborately decorated designer jeans by Ed Hardy and

The author's mother, Mary D. Shuker, wearing a dragon-decorated kimono during the 1940s. (Dr. Karl Shuker)

others, plus innumerable rings, necklaces, bracelets, and even body piercing regalia.

Moreover, the dragon's image and even its name have been incorporated into countless logos and brands—and not just in fashion. The logo of Brand Hong Kong, no less, which is a symbol utilised in the promotion of Hong Kong as an international brand name, is a dragon's head. And in contemporary Chinese culture, it is considered extremely disrespectful to deface any image of a dragon.

An offshoot of dragon-related fashion that subsequently evolved into a major field

of collectabilia is net-suke. These miniature Japanese sculptures originated during the 1600s as toggles that secured to the sash (obi) of traditional pocketless Japanese robes the cords from containers used as substitute pockets.

Originally carved from ivory but later utilising a range of other hard materials, netsuke portrayed all manner of subjects, but intricately coiled dragons were and still are a particular favourite, and can command very considerable sums of money as collectables.

DRAGON FLAGS AND KITES

During the late Qing Dynasty (which ended in 1912), the dragon appeared on China's national flag. Today, however, it is replaced by five stars, but a dragon can still be found on the flags of two other nations—Bhutan and Wales. The Bhutanese thunder dragon or druk is a typical Oriental dragon, whereas Wales's version is the red classical dragon with wings long associated with this Celtic principality within the United

Mid-19th-Century example of netsuke in the form of five intertwined Japanese dragons

Kingdom. Also worth noting is that Bhutan's national anthem is 'Druk Tsendhen'—translating as 'The Thunder Dragon Kingdom.'

Staying with dragon-inspired objects of the airborne variety: dragon kites have been manufactured for centuries, but remain very popular even in today's high-tech, gadget-driven leisure world. There are two fundamental types—figure kites, i.e. kites that mimic the form of an animal (or some object), in this instance a dragon (generally an Oriental one); and train kites, in which the dragon's body consists of a series of kites or mini-kites linked together, like carriages in a train.

DRAGONS IN GAMES AND SPORT

Kite flying can be solo or competitive, but it is not the only game or sport to feature dragons. Perhaps the best-known example globally is the fantasy role-playing game *Dungeons and Dragons* (or *D&D*, as it is known to its many millions of aficionados worldwide). Widely acclaimed as the first modern-day role-playing game, it has since spawned an entire industry of derivative versions as well as a wide range of associated branded products, but was originally designed by Gary Gygax and Dave Arneson, and was first produced in 1974. Refereed by a Dungeon Master who also serves as its storyteller and assumes the role of the world's inhabitants too, the game features players who each take the part of a specific character.

These characters interact with one another (usually as the members of a party or expedition) and the world's inhabitants when they undergo some form of quest, during which battles, trials, and searches for treasure or knowledge are integral components of the game, and feature many different types of dragon. *D&D* later inspired a same-named cartoon television series and movie. Other role-playing games involving dragons are *Earthdawn*, *Rifts*, and *Shadowrun*.

Dragons also feature heavily in fantasy-based card and trading card games. In *Magic: The Gathering*, the dragon is a specific subtype, and is represented by several important characters, such as Draco, Rorix Bladewing, the Primeval Dragons, and the Guardian Ryuu of Kamigawa. The *World of Warcraft* trading card game includes Deathwing, Onyxia, and other dragons from the very popular Real Time Strategy game series *Warcraft*. And in the trading card game *Yu-Gi-Oh!*, the dragon is a principal monster type, represented by numerous forms (e.g. sky, five-headed, ancient fairy, chaos emperor, stardust) and colours (rainbow, blue eyes white, black-winged, red eyes darkness, etc).

Although board and card games remain very popular today, the arrival of electronic games—arcade, video, and computer—changed the face of entertainment forever, and these include some famous dragon-related examples.

Undoubtedly the most successful dragon-based arcade game was *Dragon's Lair* (1983). Utilising the laserdisc video format and its massive storage potential, this enabled it to make extensive use of

Dragon boat of the Lapátolók racing team on the River Danube,
Budapest, Hungary, 10 July 2010. (Lajos Rozsa/Wikipedia)

top-quality animation supplied by ex-Disney animator Don Bluth, who had just established his own animation studio. The game's basic plot consisted of a somewhat clumsy and decidedly less-than-heroic knight, Dick the Daring, attempting to rescue the slumbering Princess Daphne from the wicked clutches of a villainous dragon called Singe, who has imprisoned her within the castle of the vile wizard Mordroc.

Over the years, there have been many video and computer games featuring dragons either as the player character or as a principal antagonist of the player character, for all of the major platforms. The earliest examples, available for the Atari 2600 console, were *Adventure* (1979) and *Dragonstomper* (1982), along with the mega-successful game *The Hobbit* (1982), available for a wide range of consoles including Amstrad CPC, Apple II, BBC Micro, Commodore 64, Sinclair ZX Spectrum, and (appropriately!) Dragon 32.

Several notable multi-console games were released during the late 1980s, such

as *Dragon Spirit* (1987), *Black Lamp* (1988), *Dragon Breed* (1989), *Saint Dragon* (1989), and *Legend of the Red Dragon* (1989). The 1990s heralded the appearance of multi-game series starring dragons, such as *Breath of Fire* (1993-2002), *Spyro* (1998-present day), and *Drakan* (1999, 2002). In 2003, an all-new, highly sophisticated *The Hobbit* video game was released, a platform-based game playable on Game Boy Advance, GameCube, Play Station 2, Windows, and Xbox, which contained several characters and 12 different chapters, each filled with tasks to accomplish and closely following events in Tolkien's original novel.

During the past decade, several prominent dragon-inspired role-playing video games have been released, and are readily available on Xbox, Windows, and Play Station. Noteworthy examples in which the dragons are antagonists include *7th Dragon* (2009), *The Elder Scrolls V: Skyrim* (2011), and *Dragon's Dogma* (2012), whereas dragons are the players in *Istaria* (2003),

I of the Dragon (2004), and *Hoard* (2010). In *Dragon Age: Origins* (2009), dragons feature both as players and as antagonists.

As for dragon sports, the best known is dragon boat racing, which originated in China as a ceremonial, religious ritual, but now takes place in competition form in many parts of the world outside the Far East, including Europe, the Middle East, Africa, North America, the Caribbean, Latin America, and Australasia. Decorated with ornate dragon heads and tails, these long wooden boats are powered by rival teams of oarsmen with paddles. Each team normally consists of a 22-man crew, 20 of which are paired paddlers, plus a drummer to maintain the paddle rhythm and a sweep who steers. The race distance ranges from 656 ft to 6560 ft, depending upon the event.

THE WORLD'S LARGEST WALKING ROBOT IS . . . A DRAGON!

Robocop may be movie-land science fiction, but a certain gigantic robot dragon is very real. Moreover, as certified by the 2014 edition of *Guinness World Records* (formerly *The Guinness Book of Records*), it is a bona fide record-holder, as the world's largest walking robot.

Conceived by a 15-strong research team in February 2007 at Zollner Elektronik AG of Germany, who then went on to design and construct it, this mechanical behemoth measures 51 ft long, weighs 11 tons, and is powered by a 0.5-gallon diesel engine boasting approximately 140 horsepower. In addition, it contains 21 gallons of stage blood, exhales real fire (using liquid gas stored beneath its polyurethane and glass-reinforced skin), is radio-controlled, and has been formally dubbed Tradinno (a combination of the words 'tradition' and 'innovation').

Possessing four sturdy legs, plus a pair of wings spanning 40 ft, Tradinno is a classical Western dragon in outward form, but whose internal system deftly combines electronic and hydraulic components, and incorporates no fewer than 238 separate sensors. These ensure that it can accurately ascertain its surrounding environment when walking, with each of its legs exhibiting seven degrees of freedom, yielding a remarkably lifelike gait.

This monstrous marvel was actually created to appear in 2010 within the *Drachenstich* (*Slaying the Dragon*), which is a 500-year-old play staged each August in the Bavarian Forest town of Furth im Wald, and it has duly starred in it each year since then. During this play, re-enacting a traditional local legend of how the town was reputedly besieged by a dragon during the Middle Ages, Tradinno spews out blood in an ostensibly miraculous manner when slain; but thanks to its newly-awarded record-holding status, this radio-controlled mega-reptilian is nowadays justly viewed as a veritable wonder in its own right too.

DRAGON COLLECTABILIA AND MEMORABILIA

Few people at some time in their lives have not succumbed to the compulsion to collect something, whether it be stamps, coins, cigarette cards, autographs, records, books,

Bronze-cast dragon
pendant. (Markus Bühler)

Intricately carved wooden lampshade
stand in the form of an Oriental water
dragon. (Dr. Karl Shuker)

or any number of much more esoteric objects. As might be expected, therefore, in view of its pre-eminence in every other field of human interest, the dragon has attracted considerable interest among collectors, and all manner of dragon-inspired collectabilia and memorabilia have been produced, especially in modern times.

One of the most popular themes is the limited-edition dragon figurine, and many different sets have been created. One of the most sought-after series, on account of the figurines' technical superiority and imaginative morphological range, was *Enchantica*. Sculpted from porcelain and resin, and also featuring wizards and witches, but mostly dragons, the series was originally manufactured by Holland Studio Craft, and the figurines began as characters from the first, bestselling *Enchantica* fantasy novel by Andrew Bill entitled *Wrath of the Ice Sorcerer*. An *Enchantica* club was also established, which lasted from 1990 to 2001.

'Garden Ware'—a delightful example of fantasy artwork depicting a flower-camouflaged garden dragon. (Pat Burroughs)

Inspired by a limited set of drawings by artist-sculptor Real Musgrave depicting a small dragon in the pocket of a tweed sports jacket produced back in the mid-1970s, *Pocket Dragons* is a highly-collectable series of figurines and other ornaments manufactured by Collectible World Studios and again designed by Musgrave. The first set, consisting of 27 figurines, was released in 1989, with over 400 different figurines available by 2006, the year in which their manufacturer went into receivership. In 1998, a cartoon television series, *Pocket Dragon Adventures*, based upon the figurine characters, was produced, and ran for 52 episodes.

Attaining frenzied levels of collectability during the late 1990s due to brilliant marketing by their manufacturer Ty (Warner) Inc, even the Beanie Babies—those small but once highly-prized plush toys stuffed with plastic beans—featured various dragons of the winged classical type among their considerable number. Scorch was greyish with bright red and yellow wings, whereas Legend was purple with brown wings, dorsal crest, and underparts, and Magic was white with silver wings. Others included striped Zodiac, gold Tempest, and even a Komodo dragon called Bali. Welsh dragons of various colours were also produced in the Beanie series. Dwynwen was violet, Dewi Y Ddraig was green, and Y Ddraig Goch was the traditional red version, Another set of plush toy dragons was the Melissa and Doug Luster series, and the larger Aurora plush dragons would roar when squeezed.

With the continuing popularity of New Age culture and related collectabilia, pewter dragons, dragons intricately carved from wood, and, for those with bigger spending budgets, dragons created from crystals, are also very much in demand.

They have inspired many different series of figurines, as well as countless dragon-featuring objects, such as candlestick holders, trinket boxes and caskets, paper knives, ceremonial daggers and swords, goblets,

Ornate Oriental dragon portrayed upon an antique umbrella stand, alongside a more recent Oriental dragon figurine carved from wood. (Dr. Karl Shuker)

picture frames, pendants, finely-carved stools, and even chairs and tables.

As for other examples: browsing through the dragon items listed under Collectables on any major internet auction site swiftly reveals that their variety is almost limitless, as is their quantity, and their costs range from pocket money prices to decidedly serious sums.

From posters, sweat shirts, pencil cases, baseball caps, and numerous other examples of dragon-related film and TV tie-in memorabilia to tea pots, bookends, fantasy artwork, mobiles, and snuff boxes, from ashtrays, chess sets, incense burners, garden ornaments, and cigarette lighters, to toothpick holders, paperweights, lamp stands, amulets, and mirrors—and everything else that could possibly be adorned in some way by the image of dragons, these reptilian monsters mesmerise humanity like no other beast, real or imagined, and seem destined to do so forever more. But why?

The continuing popularity of dragons in books, art, music, on screen, and in every other conceivable facet of our culture confirms that even though we may no longer believe in them, we certainly cannot forget them! Moreover, there may even be a very real, deep-rooted reason for their tenacity within our lives—could it be that dragons are less of a myth and more of a memory?

CHAPTER 8:
DRAGONS IN ALL OF US?

THE CULT OF THE DRAGON is as potent and vibrant today as it has always been. But why is this so? Why exactly *are* we so captivated—indeed, obsessed—by dragons? Is there something inherent and very ancient, buried deep within all of us, that explains this extraordinary anomaly—in fact, could our continuing fascination with the dragon be fuelled by prehistoric persistence?

DACQUÉ, DRAGONS, AND DINOSAURS

The first scientist to give serious credence to this remarkable possibility was German palaeontologist and Theosophist Prof. Edgar Dacqué (1878-1945). He proposed that our legends of dragons and our abiding infatuation with these huge, monstrous reptiles are of much more ancient origin than mere centuries or even millennia.

Boldly, Dacqué suggested that humans actually co-existed with dinosaurs way back in prehistoric times, and that

preserved, ancestral memories of when they lived in fear of those great reptiles were ultimately transmuted down through countless generations spanning millions of years until they re-emerged in modern-day *Homo sapiens* as a primitive but passionate belief in, and obsession with, dragons. This would therefore explain why dragons are so comparable in basic form and activity to the long-vanished dinosaurs of prehistory.

Dacentrurus, a real but very dragon-like dinosaur. (Tim Morris)

Dacqué's belief that humans and dinosaurs co-existed in prehistoric times has since been soundly disproved by the fossil record. However, his linking of dragons with ancient preserved memories of dinosaurs (some of which, like the stegosaur *Dacentrurus*, were extremely dragon-like in appearance, as revealed by fossils) is fascinating and very tantalising—but is there any tangible evidence to support it?

THE TRIUNE BRAIN
AND THE R-COMPLEX

During the 1960s, American neuroscientist and physician Dr. Paul D. MacLean formulated a revolutionary new model to explain the vertebrate brain's evolution, and called it the triune brain. He proposed that during evolution, three regions had been successively added to the forebrain. The first was the reptilian or r-complex, followed by the paleomammalian or limbic system, and then, most recently, by the neomammalian cortex or neocortex.

Relevant to Dacqué's theory was the reptilian brain, consisting of the basal ganglia, which are a group of nuclei situated at the base of the forebrain. They are intimately connected with the cerebral cortex, thalamus, and certain other brain areas, and are associated with voluntary motor control, procedural learning relating to routine behaviour, and cognitive emotional functions. MacLean claimed that the reptilian brain was the principal region of the forebrain in reptiles and birds, and was responsible for instinctive behaviour involved in ritualistic and territorial displays,

aggression, and dominance, as fully propounded in his book *The Triune Brain in Evolution* (1990).

DRAGONS OF EDEN AND
DRAGON DREAMING

More than a decade before MacLean's book was published, however, his triune brain model had already attracted particular interest from American astrophysicist and science populariser Prof. Carl Sagan, who explored and expanded his theory in a Pulitzer Prize-winning book, *The Dragons of Eden* (1978). This ground-breaking book explored many different aspects of human intelligence and the evolution of the human brain, and earned its title from Sagan's detailed contemplation of how modern-day humanity's cultural beliefs, folklore, and legends concerning snakes and dragons may derive from an ancient fear of reptiles stemming in turn from our far-distant ancestors' struggle for survival in the face of predators.

This line of thought echoed Dacqué's theory, and was further elaborated in British biologist Dr. Lyall Watson's thought-provoking book *The Dreams of Dragons* (1987). In accordance with palaeontological evidence (and therefore unlike Dacqué), however, Sagan and Watson theorised that memories of those fears and struggles may have been preserved and perpetuated from the early mammals sharing our planet with the ruling reptiles, and that they could have been retained within the r-complex of our forebrain. In humans, this is suppressed during consciousness, but reactivates and reasserts itself during sleep, populating our dreams with, albeit often in symbolic form, the veritable dragons from our prehistoric past.

DRAGONS—MYTHS OR MEMORIES?

In more recent times, MacLean's triune brain model has experienced various refutations, based upon new research by others. The most significant discovery appertaining to the r-complex is that the basal ganglia occupy a much smaller proportion of the forebrain of reptiles and birds than

The dragon—more than a myth, a Mesozoic memory? (Rebekah Sisk)

Are our legends of dragons and dragon-slayers the preserved, distorted memories of ancestral mammals' struggles with the dinosaurs?

originally supposed, and are also found in other vertebrates. Presumably, then, they did not originate with the reptiles after all, but with the vertebrates' common ancestor, dating back over 500 million years ago.

How significantly this affects the theory that our enduring infatuation with dragons—and, indeed, with dinosaurs too, judging from the plethora of dinosaur films, books, toys, and other modern-day representations—is atavistic remains contentious. Yet the parallels, morphologically as well as behaviourally, between dragons and dinosaurs are so marked that something more than mere coincidence must surely be present, planted deep and dark within our species' evolutionary heritage.

Indeed, within his compelling book *An Instinct For Dragons* (2002), anthropologist Dr. David E. Jones offers a persuasive thesis for our fear and fascination regarding dragons being a direct result of the predators that threatened our early and continuing evolution—to such an extent, in

A sky dragon, or a pterosaur from the twilit realm of imperfect
but long-inherited remembrance? (Thomas Finley)

fact, that even today our species quite literally remains 'hardwired' to believe in
dragons.

So let us end this discussion, and, indeed, this entire book, by revisiting the
notions of Dacqué, Sagan, Watson, and
Jones one last time, to reflect upon just how
likely, or otherwise, their radical proposi-
tion really is.

Back when the mighty dinosaurs, winged
pterosaurs, and other reptilian monsters
still ruled the planet, our ancestors were

little more than humble rat-like creatures,
living in their great shadows, afraid and
wary of the giants all around them. What
an enormous psychological impact the existence of these stupendous beasts must
have exerted upon the early mammals—
enough, surely, to sear indelible memories
of ferocious reptiles into their minds.

And what if those memories were preserved through all the ages that followed,
long after the dinosaurs and their kin had
vanished, retained from one mammalian

generation to the next, becoming distorted as they passed endlessly from lineage to lineage, but with their essence—the dread, and the shadowy form, of giant reptiles—remaining intact, until they were at last inherited by the first primitive humans. What would be the result?

It would most probably be a lingering, fearful fascination with giant reptiles, subconscious minds haunted by ghosts and confused nightmares where the mighty long-demised dinosaurs still roamed, and where, amid the imperfect memories of untold ages, they may even have acquired the wings of vanished pterosaurs too, ultimately metamorphosing into ferocious, wholly fictitious creatures that never existed but which we cannot forget—mutated reptilian phantoms from the Mesozoic, or, as we call them today, dragons.

SELECTED BIBLIOGRAPHY

Anon. 1996. *Dragons: An Anthology of Verse and Prose.* London: Lorenz.

Aldrovandi, Ulisse. 1640. *Serpentum et Draconum Historiae Libri Duo.* Bononiae.

Allan, Tony. 2008. *The Mythic Bestiary.* London: Duncan Baird.

Allen, Judy, and Jeanne Griffiths. 1979. *The Book of the Dragon.* London: Orbis.

Ashman, Malcolm, and Joyce Hargreaves. 1997. *Fabulous Beasts.* London: Paper Tiger.

Ashton, John. 1890. *Curious Creatures in Zoology.* London: John C. Nimmo.

Aymer, Graeme. 2009. *Dragon Art.* Fulham: Flame Tree.

Barber, Richard. 1992. *Bestiary.* London: The Folio Society.

Barber, Richard, and Anne Riches. 1971. *A Dictionary of Fabulous Beasts.* London: Macmillan.

Barrett, Charles. 1946. *The Bunyip and Other Mythical Monsters and Legends.* Melbourne: Mail Newspapers.

Base, Graeme. 1996. *The Discovery of Dragons.* London: Michael Joseph.

Bassett, Michael G. 1982. 'Formed Stones', *Folklore and Fossils.* Cardiff: National Museum of Wales.

Biedermann, Hans. 1992. *Dictionary of Symbolism.* New York: Facts On File.

Binyon, Laurence. 1911. *The Flight of the Dragon.* London: John Murray.

Blanpied, Pamela W. 1980. *Dragons: The Modern Infestation.* Woodbridge: Boydell Press.

Bölsche, Wilhelm. 1929. *Drachen: Sagen und Naturwissenschaft.* Stuttgart: Franckh'sche Verlagshandlung.

Borges, Jorge L., and Margarita Guerrero. 1974. *The Book of Imaginary Beings* Rev. edition. Harmondsworth: Penguin.

Bose, Hampden C. du. 1899. *The Dragon, Image and Demon.* Richmond: Presbyterian Committee of Publications.

Boulay, R. A. 1997. *Flying Serpents and Dragons: The Story of Mankind's Reptilian Past.* Escondido: Book Tree.

Brennan, Marie. 2013. *A Natural History of Dragons: A Memoir by Lady Trent.* London: Tor Books.

Byrne, M. St. Clare, ed. 1926. *The Elizabethan Zoo.* London: Frederick Etchells & Hugh MacDonald.

Campbell, John F. 1911. *The Celtic Dragon Myth.* Edinburgh: John Grant.

Carr-Gomm, Philip, and Stephanie Carr-Gomm. 1994. *The Druid Animal Oracle.* Old Tappan: Fireside.

Carter, Frederick. 1926. *The Dragon of the Alchemists.* London: E. Matthews.

Cherry, John, ed. 1995. *Mythical Beasts.* London: British Museum Press.

Chetwynd, Tom. 1982. *A Dictionary of Symbols.* London: Paladin.

Chetwynd, Tom. 1986. *A Dictionary of Sacred Myth.* London: Unwin.

Clair, Colin. 1967. *Unnatural History: An Illustrated Bestiary.* London: Abelard-Schuman.

Coffin, Tristram P., cons. 1984. *The Enchanted World: Dragons.* Amsterdam: Time-Life: .

Coffin, Tristram P., cons. 1985. *The Enchanted World: Magical Beasts.* Amsterdam: Time-Life.

Conway, D. J. 1996. *Magickal, Mythical, Mystical Beasts: How To Invite Them Into Your Life.* St. Paul: Llewellyn.

Cooper, J. C. 1992. *Symbolic and Mythological Animals.* London: Aquarian Press.

Costello, Peter. 1979. *The Magic Zoo: The Natural History of Fabulous Animals.* London: Sphere.

Cottrell, Annette B. 1962. *Dragons.* Boston: Museum of Fine Arts, Boston.

Crane, Pamela A. F. 1987. *Draconic Astrology.* Wellingborough: Aquarian Press.

Dance, S. Peter. 1976. *Animal Fakes and Frauds.* Maidenhead: Sampson Low.

Dell, Christopher. 2010. *Monsters: A Bestiary of the Bizarre.* London: Thames & Hudson.

Dickinson, Peter. 1979. *The Flight of Dragons.* London: Pierrot Publishing.

Dimmick, Adrian N. 1994. *Worme Worlde: The Dragon Trivia Source Book.* London: The Dragon Trust.

Dumont, Louis. 1951. *La Tarasque.* Paris: Gallimard.

Eberhart, George M. 2002. *Mysterious Creatures: A Guide to Cryptozoology.* 2 vols. Santa Barbara: ABC-Clio.

Elliot-Smith, Grafton. 1919. *The Evolution of the Dragon.* Manchester: University Press.

Evans, Jonathan. 2008. *Dragons: Myth and Legend.* London: Apple Press.

Flett, Josie. 1999. *A History of Bunyips: Australia's Great Mystery Water Beasts.* Free Spirit Press: Tyalgum.

Fox, David. 1983. *Saint George: The Saint With Three Faces.* Shooter's Lodge: Kensal.

Freeman, Richard. 2005. *Dragons: More Than a Myth?* Bideford: CFZ Press.

Freeman, Richard. 2006. *Explore Dragons.* Wymeswold: Explore Books.

Gesner, Conrad. 1551, 1554, 1555, 1558. *Historiae Animalium.* 4 vols. Zurich: Christoph Froschauer.

Gould, Charles. [Smith, Malcolm, ed.]. 1977. *The Dragon.* London: Wildwood House.

Green, Michael. 1996. *The Book of the Dragontooth.* Philadelphia: Running Press.

Green, Roger L., ed. 1970. *A Cavalcade of Dragons.* New York: H. Z. Walck.

Griffiths, Bill. 1996. *Meet the Dragon: An Introduction to Beowulf's Adversary.* Wymeswold: Heart of Albion Press.

Gubernatis, Angelo de. 1872. *Zoological Mythology: or, The Legends of Animals* 2 vols. London: Trübner.

Hardy, Donald E. 1988. *Dragon Tattoo Design.* San Francisco: Hardy Marks.

Hargreaves, Joyce. 1983. *The Dragon Hunter's Handbook.* London: Granada.

Hargreaves, Joyce. 1990. *Hargreaves New Illustrated Bestiary.* Glastonbury: Gothic Image.

Hayes, L. Newton. 1922. *The Chinese Dragon.* Commercial Press: Shanghai.

Hesilrige, Arthur G. M., ed. n.d. *Debrett's Heraldry.* London: Dean & Son.

Heuvelmans, Bernard. 1958. *On the Track of Unknown Animals.* London: Rupert Hart-Davis.

Heuvelmans, Bernard. 1968. *In the Wake of the Sea-Serpents.* London: Rupert Hart-Davis.

Hogarth, Peter, and Val Clery. 1979. *Dragons.* London: Allen Lane.

Hoke, Helen, ed. 1972. *Dragons, Dragons, Dragons.* New York: Franklin Watts.

Holiday, F. W. 1963. *The Dragon and the Disc.* London: Sidgwick & Jackson.

Holman, Felice, and Nanine Valen. 1975. *The Drac: French Tales of Dragons and Demons.* New York: Charles Scribner's Sons.

Hoult, Janet. 1987. *Dragons: Their History and Symbolism.* Glastonbury: Gothic Image.

Huber, Richard. 1981. *Treasury of Fantastic and Mythological Creatures.* New York: Dover.

Hulme, F. Edward. 1895. *Natural History Lore and Legend.* London: B. Quaritch.

Huxley, Francis. 1979. *The Dragon: Nature of Spirit, Spirit of Nature.* London: Thames & Hudson.

Ingersoll, Ernest. 1928. *Dragons and Dragon Lore.* New York: Payson & Clarke.

Johnsgard, Paul, and Karin Johnsgard. 1982. *Dragons and Unicorns: A Natural History.* New York: St. Martin's Press.

Jones, David E. 2002. *An Instinct For Dragons.* New York: Routledge.

Lehner, Ernst, and Johanna Lehner. 1969. *A Fantastic Bestiary: Beasts and Monsters in Myth and Folklore* (New York: Tudor); republished as Lehner, Ernst, and Johanna Lehner. 2004. *Big Book of Dragons, Monsters, and Other Mythical Creatures.* Mineola: Dover.

Lofmark, Carl. 1995. *A History of the Red Dragon.* Llanrwst. Gwasg Carreg Gwalch.

Lum, Peter. 1952. *Fabulous Beasts.* London: Thames & Hudson.

Macdonald, Fiona. [Consultant: Shuker, Karl P. N.] 2001. *Monsters.* London: Lorenz.

Mackal, Roy P. 1987. *A Living Dinosaur? In Search of Mokele-Mbembe.* Leiden: E. J. Brill.

MacLean, Paul D. 1990. *The Triune Brain in Evolution.* New York: Plenum.

Matthews, John, and Caitlin Matthews. 2005. *The Element Encyclopedia of Magical Creatures.* London: HarperElement.

Meurger, Michel. 2001. *Histoire Naturelle des Dragons.* Rennes: Terre de Brume.

Newman, Paul. 1980. *The Hill of the Dragon: An Enquiry Into the Nature of*

Dragon Legends. Ottowa: Rowman & Littlefield.

Newton, Michael. 2005. *Encyclopedia of Cryptozoology: A Global Guide.* Jefferson, NC: McFarland.

Nigg, Joseph, ed. 1999. *The Book of Fabulous Beasts: A Treasury of Writings From Ancient Times to the Present.* Oxford: Oxford University Press.

Noble, Marty. 2002. *Dragons: A Book of Designs.* New York: Dover.

O'Connell, Mark, et al. 2010. *The Ultimate Illustrated Encyclopedia of Signs, Symbols and Dream Interpretation.* London: Hermes House.

Paciorek, Andrew L. 2011. *Strange Lands: A Field-Guide to the Celtic Otherworld.* http://www.blurb.co.uk/b/1957828-strange-lands

Page, Michael, and Robert Ingpen. 1985. *Encyclopaedia of Things That Never Were.* Limpsfield: Lansdowne Press/Dragon's World.

Passes, David. 1993. *Dragons: Truth, Myth and Legend.* St. Albans: David Bennett.

Pennick, Nigel. 1997. *Dragons of the West.* Chieveley: Capall Bann.

Phillips, Henry. 1882. *Basilisks and Cockatrices.* Philadelphia: E. Stern.

Pickering, Fran. 1999. *The Element Illustrated Encyclopedia of Animals in Nature, Myth and Spirit.* London: Element.

Rosen, Brenda. 2010. *The Mythical Creatures Bible: The Definitive Guide to Beasts and Beings From Mythology and Folklore.* London: Godsfield Press.

Rowland, Beryl. 1973. *Animals With Human Faces: A Guide to Animal Symbolism.* Knoxville: University of Tennessee Press.

Rudd, Elizabeth, ed. 1980. *Dragons.* London: W. H. Allen.

Sagan, Carl. 1978. *The Dragons of Eden.* London: Hodder & Stoughton.

Salverte, Eusebe B. de. 1826. *Des Dragons et des Serpents Monstrueux qui Figurent dans un Grand Nombre de Récits Fabuleux ou Historiques.* Paris: Rignoux.

Sanders, Malcolm. 2002. *The Dragon Chronicles.* Limpsfield: Pegasus Publishing.

Sanders, Tao Tao Liu. 1983. *Dragons, Gods and Spirits From Chinese Mythology.* New York: Schocken.

Screeton, Paul. 1978. *The Lambton Worm and Other Northumbrian Dragon Legends.* Fulham: Zodiac House.

Screeton, Paul. 1998. *Whisht Lads and Haad Yor Gobs: The Lambton Worm and Other Northumberland Dragon Legends.* Pennywell: Northeast Press.

Sedgwick, Paulita. 1974. *Mythological Creatures: A Pictorial Dictionary.* New York: Holt, Rinehart & Winston.

Shuker, Karl P. N. 1995. *Dragons: A Natural History.* New York: Simon & Schuster/London: Aurum Press.

Shuker, Karl P. N. 1995. *In Search of Prehistoric Survivors.* London: Blandford Press.

Shuker, Karl P. N. 1997. *From Flying Toads To Snakes With Wings.* St. Paul: Llewellyn.

Shuker, Karl P. N. 2003. *The Beasts That Hide From Man.* New York: Paraview.

Shuker, Karl P. N. 2013. *Mirabilis: A Carnival of Cryptozoology and Unnatural*

History. New York: Anomalist Books.

Simpson, Jacqueline. 1980. *British Dragons*. London: B.T. Batsford.

South, Malcolm, ed. 1981. *Topsell's Histories of Beasts*. Chicago: Nelson-Hall.

Steer, Dugald A., ed. 2009. *Drake's Comprehensive Compendium of Dragonology*. Dorking: Templar Publishing.

Summers, Lori. 2002. *The Dragon Hunter's Handbook*. London: Penguin.

Topsell, Edward. 1607. *The Historie of Foure-Footed Beastes*. London: William Iaggard.

Topsell, Edward. 1608. *The Historie of Serpents*. London: William Iaggard.

Trubshaw, Bob. 1993. *Dragon Slaying Myths Ancient and Modern*. Wymeswold: Heart of Albion Press.

Vinycomb, John. 1906. *Fictitious and Symbolic Creatures in Art*. London: Chapman & Hall.

Visser, Marinus de. 1858. *The Dragon in China and Japan*. Amsterdam: Müller.

Waddell, Helen, trans. 1934. *Beasts and Saints*. London: Burns Oates & Washbourne.

Watkins, M. G. 1885. *Gleanings From the Natural History of the Ancients*. London: Elliot Stock.

Watson, Lyall. 1987. *The Dreams of Dragons*. Scranton: William Morrow.

Wharton, Violet. 1994. *Dragons and Fabulous Beasts*. London: Pavilion.

White, T. H. 1954. *The Book of Beasts*. London: Jonathan Cape.

Whitlock, Ralph. 1983. *Here Be Dragons*. London: George Allen & Unwin.

Woodruff, Una. 1979. *Inventorum Natura*. London: Dragon's World.

Zell-Ravenheart, Oberon, and Ash L. DeKirk. 2007. *A Wizard's Bestiary*. Franklin Lakes: New Page.

INDEX

AUTHOR BIOGRAPHY

BORN AND STILL LIVING in the West Midlands, England, Dr. Karl P.N. Shuker graduated from the University of Leeds with a Bachelor of Science (Honours) degree in pure zoology, and from the University of Birmingham with a Doctor of Philosophy degree in zoology and comparative physiology. He now works full-time as a freelance zoological consultant to the media, and as a prolific published writer.

Dr. Shuker is currently the author of 20 books and hundreds of articles, principally on animal-related subjects, with an especial interest in cryptozoology and animal mythology, on which he is an internationally-recognised authority, but also including a poetry volume. In addition, he has acted as consultant for several major multi-contributor volumes as well as for the world-renowned *Guinness Book of Records /Guinness World Records* (he is currently its Senior Consultant for its Life Sciences section); and he has compiled questions for the BBC's long-running cerebral quiz 'Mastermind.' He is also the editor of the *Journal of Cryptozoology*, the world's only existing peer-reviewed scientific journal devoted to mystery animals.

Dr. Shuker has travelled the world in the course of his researches and writings, and has appeared regularly on television and radio. Aside from work, his diverse range of interests include motorbikes, the life and career of James Dean, collecting masquerade and carnival masks, quizzes, philately, poetry, travel, world mythology, and the history of animation.

He is a Scientific Fellow of the prestigious Zoological Society of London, and a Fellow of the Royal Entomological Society. He is Cryptozoology Consultant to the Centre for Fortean Zoology, and is also a Member of the Society of Authors.

Dr. Shuker's personal website can be accessed at http://www.karlshuker.com and his mystery animals blog, ShukerNature, can be accessed at http://www.karlshuker. blogspot.com

His poetry blog can be accessed at http:// starsteeds.blogspot.com and his Eclectarium blog can be accessed at http:// eclectariumshuker.blogspot.com

The author alongside his wood-carved Oriental dragon statue (Dr. Karl Shuker)

There is also an entry for Dr. Shuker in the online encyclopedia Wikipedia at http://en.wikipedia.org/wiki/Karl_Shuker and a Like (fan) page on Facebook.

AUTHOR BIBLIOGRAPHY

1989. *Mystery Cats of the World: From Blue Tigers To Exmoor Beasts.* London: Robert Hale.

1991. *Extraordinary Animals Worldwide.* London: Robert Hale.

1993. *The Lost Ark: New and Rediscovered Animals of the 20th Century.* London: HarperCollins.

1995. *Dragons: A Natural History.* London: Aurum/New York: Simon & Schuster; republished 2006, Cologne: Taschen.

1995. *In Search of Prehistoric Survivors: Do Giant 'Extinct' Creatures Still Exist?* London: Blandford.

1996. *The Unexplained: An Illustrated Guide to the World's Natural and Paranormal Mysteries.* London: Carlton/North Dighton: JG Press; republished 2002, London: Carlton.

1997. *From Flying Toads To Snakes With Wings: From the Pages of FATE Magazine.* St. Paul: Llewellyn; republished 2005, London: Bounty.

1999. *Mysteries of Planet Earth: An Encyclopedia of the Inexplicable.* London: Carlton.

2001. *The Hidden Powers of Animals: Uncovering the Secrets of Nature* (Pleasantville: Reader's Digest/London: Marshall Editions.

2002. *The New Zoo: New and Rediscovered Animals of the Twentieth Century* [fully-updated, greatly-expanded, second edition of *The Lost Ark*]. (Thirsk, UK: House of Stratus Ltd./Poughkeep-sie, NY: House of Stratus Inc.

2003. *The Beasts That Hide From Man: Seeking the World's Last Undiscovered Animals.* New York: Paraview.

2007. *Extraordinary Animals Revisited: From Singing Dogs To Serpent Kings.* Bideford: CFZ Press.

2008. *Dr. Shuker's Casebook: In Pursuit of Marvels and Mysteries.* Bideford: CFZ Press.

2008. *Dinosaurs and Other Prehistoric Animals on Stamps: A Worldwide Catalogue.* Bideford: CFZ Press.

2009. *Star Steeds and Other Dreams: The Collected Poems.* Bideford: CFZ Press.

2010. *Karl Shuker's Alien Zoo: From the Pages of Fortean Times.* Bideford: CFZ Press.

2012. *The Encyclopaedia of New and Rediscovered Animals: From The Lost Ark to The New Zoo—and Beyond* [fully-updated, greatly—expanded, third edition of *The Lost Ark*]. Landisville, PA: Coachwhip Publications.

2012. *Cats of Magic, Mythology, and Mystery: A Feline Phantasmagoria.* Bideford: CFZ Press.

2013. *Mirabilis: A Carnival of Cryptozoology and Unnatural History.* New York: Anomalist Books.

2013. *Dragons in Zoology, Cryptozoology, and Culture.* Greenville, OH: Coachwhip Publications.

CONSULTANT AND ALSO CONTRIBUTOR

1993. *Man and Beast*. Pleasantville, New York: Reader's Digest.

1993. *Secrets of the Natural World*. Pleasantville, New York: Reader's Digest.

1995. *Almanac of the Uncanny*. Surry Hills, Australia: Reader's Digest.

1997-present day. *The Guinness Book of Records/Guinness World Records*. 1998-present day. London: Guinness.

CONSULTANT

2001. *Monsters*. London: Lorenz.

CONTRIBUTOR

1995. *Of Monsters and Miracles* CD-ROM. Oxton: Croydon Museum/Interactive Designs.

1996. *Fortean Times Weird Year 1996*. London: John Brown Publishing.

1998. *Mysteries of the Deep*. St. Paul: Llewellyn.

1999. *Guinness Amazing Future*. London: Guinness.

2000. *The Earth*. London: Channel 4 Books.

2001. *Mysteries and Monsters of the Sea*. New York: Gramercy.

2007. *Chambers Dictionary of the Unexplained*. Edinburgh: Chambers.

2008. *Chambers Myths and Mysteries*. Edinburgh: Chambers.

2009. *The Fortean Times Paranormal Handbook*. London: Dennis Publishing.

Plus numerous contributions to the annual *CFZ Yearbook* series of volumes.

EDITOR

Journal of Cryptozoology (published by CFZ Press).

COACHWHIP PUBLICATIONS

COACHWHIPBOOKS.COM

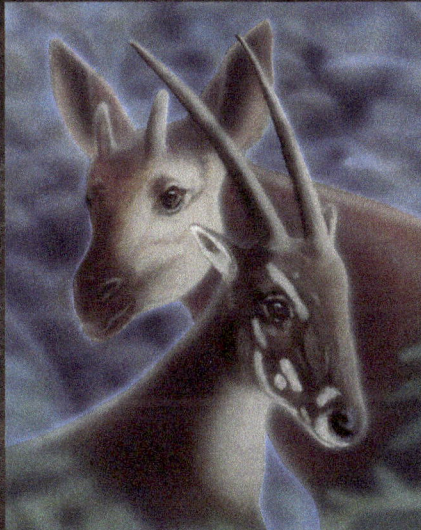

THE ENCYCLOPÆDIA OF NEW AND REDISCOVERED ANIMALS

FROM THE LOST ARK TO THE NEW ZOO—AND BEYOND

DR. KARL P.N. SHUKER

The Encyclopaedia of New and Rediscovered Animals
Dr. Karl P. N. Shuker
ISBN 978-1-61646-130-0